STP 1222

Buried Plastic Pipe Technology: 2nd Volume

Dave Eckstein, Editor

ASTM Publications Code Number (PCN):
04-012220-58

ASTM
1916 Race Street
Philadelphia, PA 19103
Printed in the U.S.A.

Library of Congress Cataloging-in-Publication Data

Buried plastic pipe technology: 2nd volume / Dave Eckstein, editor.
(Special technical publication ; 1222)
"Papers presented at the symposium ... held in New Orleans, LA from 28 Feb. to 2 March 1994" --CIP foreword.
Includes bibliographical references and index.
ISBN 0-8031-1992-5
1. Underground plastic pipe--Congresses. II. Eckstein, Dave. 1954- . II. American Society for Testing and Materials. III. Series: ASTM special technical publication ; 1222.
TJ930-B873 1994 94-10977
CIP

Peer Review Policy

Each paper published in this volume was evaluated by three peer reviewers. The authors addressed all of the reviewers' comments to the satisfaction of both the technical editor(s) and the ASTM Committee on Publications.

To make technical information available as quickly as possible, the peer-reviewed papers in this publication were printed "camera-ready" as submitted by the authors.

The quality of the papers in this publication reflects not only the obvious efforts of the authors and the technical editor(s), but also the work of these peer reviewers. The ASTM Committee on Publications acknowledges with appreciation their dedication and contribution to time and effort on behalf of ASTM.

Printed in Baltimore, MD
May 1994

Foreword

This publication, *Buried Plastic Pipe Technology: 2nd Volume,* contains papers presented at the symposium of the same name, held in New Orleans, LA from 28 Feb. to 2 March 1994. The symposium was sponsored by ASTM Committee F-17 on Plastic Piping Systems, D-20 on Plastics, and Subcommittee D20.23 on Reinforced Plastic Piping Systems and Chemical Equipment. Dave Eckstein of Uni-Bell PVC Pipe association in Dallas, TX presided as symposium chairman and is the editor of the resulting publication.

Contents

Trenchless Construction

Overview

The second symposium on Buried Plastic Pipe Technology is just what the title implies, a sequel to the first. Given the success of the first symposium, the instruction from the steering committee was brief and succinct, "Follow exactly the format from the first symposium, but ensure that the content represents state-of-the-art technical input for today." Four years having elapsed, coupled with the ever-expanding topic of buried plastic pipes facilitated accomplishing this goal.

The papers are categorized into five sections of: Field Testing, Design and Installation, Rehabilitation, Laboratory Testing, and Trenchless Construction.

Howard et al. report detailed field measurements of a 915-mm fiberglass pipe installation in the former USSR, now Latvia.

I. D. Moore introduces a three-dimensional viscoelastic finite-element model to predict circumferential stress and strain in HDPE pipes. The paper compares results with that of conventional parallel plate stiffness evaluation in predicting actual behavior. Next, A. Howard reports on the Bureau of Reclamation's 25 years of experience with soil-cement slurry pipe bedding. Critical parameters are defined and discussed.

L. J. Petroff offers a design methodology for buried HDPE manholes that accounts for both the ring-directed and axially-directed effects of applied earth pressure. Groundwater loadings and "downdrag" of surrounding soil are also investigated.

The controlled expansion of conventionally extruded PVC pressure pipe produces a preferred molecular orientation that results in increased tensile strength and other performance enhancements. D. E. Bauer reports on over a decade of field experience and research and testing with oriented PVC pipe.

Two papers provide analysis of rehabilitation techniques on two completely different aspects of their application. D. G. Kleweno reports on chemical exposures to six commercially available resins for cured-in-place pipe rehabilitation. Lo and Zhang propose two separate collapse models for encased pipes. Special attention is given to the analysis of the annular gap between the two pipes and the effects of hydrostatic loading and temperature variations.

The next section, Laboratory Testing, provides four papers on a wide range of investigated parameters. Woods and Ferry report on the phenomenon of compressive buckling of hollow cylinders during pressure testing. When the phenomenon may exhibit itself and specific recommendations for test apparatus are included.

A new test for studying behavior of buried plastic pipes in hoop compression is presented by Selig et al. A cylindrical steel vessel with an inflatable bladder serves as the core apparatus for this new test procedure.

Leevers et al. provide an extensive investigation of rapid crack propagation in polyethylene pipe materials. Several test methods and their relative ability to predict RCP in polyethylene are presented.

The effects of acid environment on PVC pipes is presented in two papers back-to-back. Sharff and DelloRusso report on a two-year study exposing PVC pipes held at a constant 5% deflection to 1.0N solution of sulfuric acid with minimal effect.

Hawkins and Mass, who begin the section on Trenchless Construction, report on results of 14-day to 6-month exposures of calcium-carbonate filled PVC pipes to 20% sulfuric acid environments. Scanning electron microscopy and wavelength dispersive x-ray microanalysis are used to provide qualitative and quantitative effects to the calcium carbonate and PVC combination.

Tohda et al. conclude a non-conservative possibility with current Japanese design standards for predicting bending moment and pipe deflection when pipes are installed open excavation using sheet piling. Centrifuge model tests used to reach this conclusion are described in detail.

McGrath et al. investigate the effect of short-term loading to a polyethylene pipe already subjected to long-term load. An example would be traffic loading on a buried pipe. The simulating test protocol is described and results reported.

The final three papers by Iseley et al., Najafi and Iseley, and Brown and Lu complete this publication. The first (perhaps more appropriately rehabilitation) categorizes and summarizes six trenchless methods as cured-in-place pipes, sliplining, in-line replacement, deformed and reshaped, point source repair, and sewer manhole rehabilitation. The second paper chronicles a full-scale test of PVC profile wall sewer pipe for microtunneling using a new microtunneling propulsion system. The final paper by Brown and Lu investigates RCP in polyethylene gas pipes specific to the effects of loading rates.

The goal of the symposium and this STP was to provide an update in the technology of buried plastic pipe. We hope you agree that we have succeeded.

I would like to extend my personal gratitude to all of those who contributed to the success of this effort but who might otherwise go unrecognized. Special thanks to the ASTM staff, the steering committee, and the many reviewers of these papers.

Dave Eckstein

Uni-Bell PVC Pipe Association
2655 Villa Creek Dr., Suite 155, Dallas, TX 75234;
symposium chairman and editor.

Field Testing

Amster Howard,[1] Juris B. Spridzans,[2] and B. J. Schrock[3]

LATVIA FIELD TEST OF 915-mm FIBERGLASS PIPE

REFERENCE: Howard, Amster, Spridzans, J. B., and Schrock, B. J., **"Latvia Field Test of 915-mm Fiberglass Pipe,"** Buried Plastic Pipe Technology: 2nd Volume, ASTM STP 1222, Dave Eckstein, Ed., American Society for Testing and Materials, Philadelphia, 1994.

ABSTRACT: The USA and USSR jointly constructed a special test section of 915-mm diameter Reinforced Plastic Mortar (RPM) fiberglass pipe in June 1979 near Riga, Latvia. This experiment was part of the working agreement of the US-USSR team "Investigations of Effectiveness of Plastic Pipe in Drainage and Irrigation." Measurements were made of pipe deflections, soil properties, and in-place densities. Six different embedment conditions were used. The pipe deflections were measured during each state of construction and over a 4-year period.

Data of particular interest is the increase in the vertical diameters caused during soil compaction at the sides of the pipe and the frequent deflection measurements in the few weeks following the final placement of the 3 m of backfill over the pipe. The ratio of the vertical deflection after 4 years to the vertical deflection on the day the backfilling was completed ranges from 1.6 to 1.7 for the dumped side support, 4.5 for a side support with a moderate degree of compaction, and 2.2 to 2.9 for the side support placed to a high degree of compaction.

KEY WORDS: pipe, fiberglass pipe, flexible pipe, deflection, test section, soil mechanics, soil tests, time factors, casper, soil-structure interaction

---

This paper reports the results of deflections measured over 4 years for 915-mm RPM (reinforced plastic mortar) pipe buried as part of a two-country joint experiment.

---

[1] Research Civil Engineer, U.S. Bureau of Reclamation, PO Box 25007, Denver CO 80225

[2] Chief, Polymer Conduits Branch, VNII Vodpolimer, 229601 Jelgava, Latvia

[3] President, B.J.S. Engineering Co., Sacramento, California

The experiment was conducted to evaluate the load deflection behavior of buried RPM pipe. The pipe was installed in 1979 with six different embedment conditions at a site near Riga, Latvia. This installation was unique in that measurements made at each increment of construction resulted in data very seldom collected. The result was increased knowledge about the change in the vertical diameters during compaction alongside the pipe, deflections related to level of backfill over the pipe, and the increase in deflection during the first few days and weeks following installation.

This experiment was part of the working agreement of the US-USSR team, "Investigation of Effectiveness of Plastic Pipe in Drainage and Irrigation." The team was part of the Soviet-American Joint Commission on Scientific and Technical Cooperation Program that was active from 1972 to 1982. This paper is a summary of a series of Bureau of Reclamation internal memorandums issued from 1979 to 1989. [1]

PIPE

The RPM pipe is 915 mm inside diameter with approximately 10-mm wall thickness and was rated at 450 feet of head. The pipe was manufactured in Riverside, California, and shipped to Latvia. Parallel plate tests on sections of the pipe showed the EI/r^3 of the pipe to be about 18 kN/m^2 (2.6 lb/in^2). The pipe to be monitored in the test reach are 5.5 m long, and the transitional pieces between the monitored sections are 2.7 m long. The pipe has bell- and spigot-type joints.

CONSTRUCTION

A typical cross section for the six bedding conditions for the RPM pipe is shown on figure 1.

The natural moisture and density of the trench wall material, a sandy clay, was determined by both a sand cone device and a densitometer-moisture gauge. The trench walls were firm, having a density of about 2.0 Mg/m^3, and had a moisture content of about 13 percent.

About a 50-mm layer of sand (same source as the sand used beside the pipe) was spread in the bottom of the trench and the bottom fine graded. Two 4-m sections of 915-mm-inside-diameter reinforced concrete pipe were placed at the downstream end of the test section which daylighted on the bank of a lake. The RPM pipe was then laid and joined. At the end of the RPM pipe section, a manhole was constructed using precast reinforced concrete rings to provide access into the pipe.

After laying the pipe, a small wedge of soil was pushed into the pipe haunch area and hand compacted to prevent sideways displacement of the pipe.

At this point, the first diameter measurements were made in the RPM pipe with an inside micrometer and deflectometer.

The deflectometer continuously recorded on a strip chart the vertical and horizontal diameter of the pipe as it was pulled through the pipe.

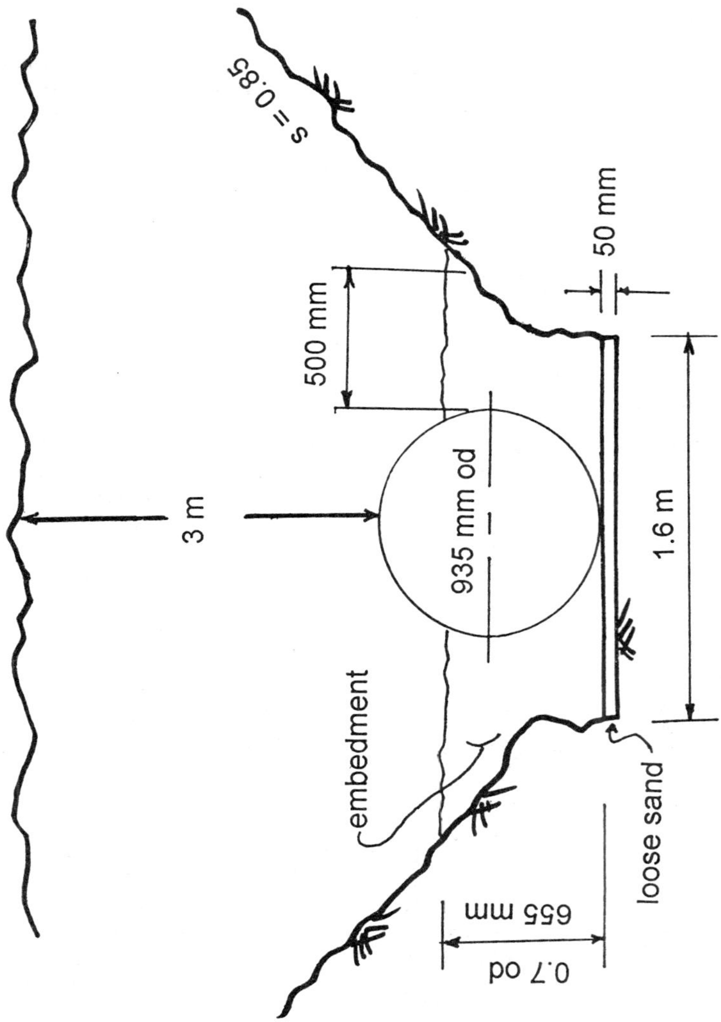

FIG. 1 - - Typical Trench Dimensions and Cross-Section

The locations of the readings were based on the bump in the tracing on the strip chart as the deflectometer went by a joint. Numerical values scaled off of the strip chart were recorded to the nearest 1 mm.

The readings taken by the two methods were within 2 mm of each other (about 0.2 percent of the pipe diameter).

The compacted bedding was constructed using four compacted layers of 165-mm thickness each to obtain a bedding height of 0.7 outside diameter. The clay was placed as its natural moisture of 10 to 13 percent, which is about optimum. The sand was placed at its natural moisture content of about 2 percent and water was added to the sand before compacting. The moistures from the field density tests of the sand ranged from 4 to 8 percent, below the optimum of about 17 percent. After completion of the four compacted bedding reaches, the dumped reaches were constructed by dumping soil beside the pipe to a height of about 0.7 o.d. using a dragline and then leveling with shovels to the 0.7 o.d. height. No water was added to the dumped sand. No foot traffic was allowed on the dumped reaches.

After each loose layer of soil had been placed in the trench and leveled, all rocks 75 mm and larger were removed by hand before compacting the layer. Compaction was done using two hand-held electric "wacker-type" tampers working simultaneously with one on each side of the pipe. By observation and measurement, it was determined that the pipe was not raised during compaction of the first layer.

Soil densities were measured in each compacted layer with a densitometer-moisture device and with the sand come device after the second, third, and fourth layers. Pipe diameter measurements were made with both the micrometer and the deflectometer after the second, third, and fourth layers were compacted. For the dumped reaches, the diameters were measured after dumping the bedding soil in, but before any backfill was placed. Densities were measured in the dumped bedding material after 0.2 to 0.3 m of backfill had been placed over the bedding.

The backfill over the pipe was placed in four lifts. First, the dragline placed the native soil from the stockpile over the pipe which was then leveled by hand to a height of 0.2 to 0.3 m over the top of the pipe. Rocks over 75 mm were removed by hand if close to the pipe. Next, a bulldozer pushed in material from the stockpile on one side of the trench, and the dragline leveled the soil so that was about 1 m of cover over the pipe. For the third layer, the bulldozer pushed in all the material needed from the stockpile and then leveled the material so there was about 2 m of cover. This same method was then used for the fourth layer resulting in a final cover of 3 m.

SOIL PROPERTIES

All of the properties of the soils were determined by USSR standards. The clay material used for bedding was the native soil excavated from the trench and was a sandy clay (CL/SC) with about 5 percent gravel and 50 percent fines with medium plasticity (minus 75-mm fraction). The liquid limit was 19 and the plasticity index was 9. From visual observations, there was about 5 to 15 percent cobbles and boulders based on the total material. The specific gravity was 2.65. The maximum dry density was 1.00 Mg/m^3 at an optimum moisture of 11 percent.

The sand bedding material came from a nearby gravel pit and was a poorly graded sand (SP). About 75 to 85 percent of the sand had particle diameters between 0.25 and 1.0 mm and had 2 to 3 percent fines (minus 75 μm) based on visual observation. The specific gravity was 2.62 to 2.59.

The maximum density was 1.68 Mg/m^3 at an optimum moisture of 16 percent. The USSR "Proctor-type" test has a higher energy input than ASTM D 698.

The in-place density of the soil placed beside the pipe was frequently measured with both the sand-cone device and the densitometer.

The degree of compaction of the six test reaches is shown in table 1.

Table 1. - Soil densities in each test reach

Test reach	Soil type	Placement	Range of densities (Percent Proctor)	Average percent Proctor	Compaction category
A	Sand	Dumped	89	89	Dumped
B	Sand	Compacted	96 to 102	98	High
C	Sand	Compacted	101 to 106	103	High
D	Clay	Dumped	76	76	Dumped
E	Clay	Compacted	95 to 98	97	High
F	Clay	Compacted	89 to 93	92	Moderate

BEDDING CONDITIONS

Six different bedding conditions were established with two different soil types used with three different degrees of compaction applied to each. For the sand bedding, two test reaches that were intended to have different degrees of compaction ended up having the same degree of compaction.

Each embedment condition was given a separate designation as shown in table 2.

Table 2. - Embedment condition

Test reach	Soil type	Degree of compaction	Percent Proctor represented by degree of compaction
A	Sand	Dumped	No compaction
B	Sand	High	Over 95 percent
C	Sand	High	Over 95 percent
D	Clay	Dumped	No compaction
E	Clay	High	Over 95 percent
F	Clay	Moderate	85 to 95 percent

DEFLECTION MEASUREMENTS

All deflections discussed are the vertical deflections of the pipe unless otherwise described. Deflection is defined as the decrease in the vertical diameter of the pipe due to the backfill soil being placed above a height equal to 0.7 of the outside diameter. Percent deflection is defined as:

$$\Delta Y\ (\%) = \frac{\text{change in diameter}}{\text{original diameter}} \times 100$$

"Change in diameter" is the diameter measured when the bedding was completed up to 0.7 of the outside diameter minus the diameter measured at some time during or after backfilling. "Original diameter" is the nominal inside diameter of the pipe, 915 mm.

Each test reach consisted of two pipe units including the joint in between. Vertical and horizontal diameter measurements were made at seven locations in each test reach - at three locations in each pipe barrel and at the joint. The average of the six readings in the pipe barrels was used to calculate the deflection for each test reach. The joint deflections are discussed separately.

PIPE OBLONGATION DURING BACKFILLING

Flexible pipe can oblongate (increase in the vertical diameter and decrease in the horizontal diameter) due to compaction of the bedding soil alongside the pipe. The diameters (horizontal and vertical) of the pipe were measured with the pipe resting in place on the trench bottom before any bedding soil was placed. The compacted bedding was placed in four lifts, each of 165 mm compacted thickness. Diameter measurements were made after the second, third, and fourth lifts. The dumped bedding was placed in one lift and diameter measurements were made after placement.

There was a linear relationship between the increase in oblongation and the height of the compacted bedding. The horizontal diameter change was larger than the vertical diameter change for the compacted beddings as summarized in table 3.

Table 3. - Oblongation due to compaction

Test reach	Soil type	Degree of compaction	Percent average diameter oblongitudinal change with bedding at 0.7 o.d.	
			Vertical	Horizontal
A	Sand	Dumped	0.3	0.3
B	Sand	High	1.3	1.4
C	Sand	High	2.0	2.2
D	Clay	Dumped	0.2	0.2
E	Clay	High	1.5	1.7
F	Clay	Moderate	1.2	1.3

The amount of oblongation was directly related to the compactive effort applied to the bedding soil. The maximum and minimum vertical oblongations are shown in table 4 along with the average vertical oblongation for all six readings in the pipe barrel

Table 4. - Range of vertical oblongations

Test reach	Soil type	Degree of compaction	Percent vertical oblongation with bedding at 0.7 o.d.		
			Max.	Min.	Avg.
A	Sand	Dumped	0.4	0.2	0.3
B	Sand	High	1.7	1.1	1.3
C	Sand	High	2.4	1.7	2.0
D	Clay	Dumped	0.2	0.1	0.2
E	Clay	High	1.9	1.3	1.5
F	Clay	Moderate	1.4	1.0	1.2

The results show that just dumping soil beside a pipe can result in oblongation. Compacting the bedding soil to over 90 percent Proctor can oblongate the pipe 1 to 2 percent.

PIPE DEFLECTION DURING BACKFILLING

Flexible pipe deflects (decreases in the vertical diameter and increases in the horizontal diameter) due to the backfill load on the pipe. The initial diameter (or zero) reading for calculating deflection was the pipe diameter measured when the bedding soil was at a height equal to 0.7 o.d. From this zero point, any changes in the pipe diameters are due to the loose clay backfill placed over the pipe. The backfill was placed in four lifts. The first lift was placed from 0.7 o.d. to a height of 0.2 to 0.3 m over the top of the pipe. The second lift was an additional 0.8 m of soil, which brought the total height of cover over the pipe to 1 m. Two additional lifts of 1 m each were placed over the pipe for a total cover height of 3 m. Diameter readings were made for all four backfill lifts.

The maximum deflection did not always occur at midspan. Therefore, the deflections discussed for the pipe for each test reach are the average of the six barrel deflection readings in each reach. Except for the dumped clay reach (reach D), the average of the six readings was within 0.1 percent of the average of the midspan readings (stations 2 and 5). The change in deflection with cover depth is illustrated in figure 2.

The following table 5 shows the maximum and minimum vertical deflections at 3 m of cover along with the average deflection.

Table 5. - Range of deflections

			Percent vertical deflection at 3 m of cover		
Test reach	Soil type	Degree of compaction	Max.	Min.	Avg.
A	Sand	Dumped	2.3	1.4	1.9
B	Sand	High	0.8	0.2	0.5
C	Sand	High	0.7	0.3	0.6
D	Clay	Dumped	9.1	6.5	7.8
E	Clay	High	1.3	0.7	1.1
F	Clay	Moderate	1.6	0.9	1.1

Vertical Versus Horizontal Diameter Changes

The horizontal deflections were smaller than the vertical deflections as summarized in table 6.

Table 6. - Average percent deflections

Test reach	Soil type	Degree of compaction	Average percent deflection		Ratio
			Vertical ΔY	Horizontal ΔX	$\Delta X/\Delta Y$
A	Sand	Dumped	1.9	1.9	1.0
B	Sand	High	0.5	0.1	0.2
C	Sand	High	0.6	0.1	0.2
D	Clay	Dumped	7.8	7.9	1.0
E	Clay	High	1.1	0.6	0.6
F	Clay	Moderate	1.1	0.6	0.6

The vertical and horizontal deflections of a flexible pipe are normally assumed to be about the same. For pipe that deflects elliptically, the ratio of the horizontal to vertical deflections is about 0.91 [4]. For test reaches A and D where the bedding soil was dumped in, the $\Delta X/\Delta Y$ ratios were 1.00 and the pipe deflected elliptically because of the poor side soil support.

For the test reaches with compacted beddings, the $\Delta X/\Delta Y$ ratios were 0.2 and 0.6 for the sand and clay soils, respectively. The compacted soil at the sides of the pipe provided enough resistance to the horizontal increase in diameter that the pipe deformed in a rectangular shape rather than an elliptical shape.

Net Change in Pipe Diameter

The net change in the pipe diameter from the measurements made when the pipe was in place on the trench bottom and after backfilling was completed is shown in table 7.

Table 7. - Net change in pipe diameter

Test reach	Soil type	Degree of compaction	Oblongation percent	Deflection percent	Net change percent
A	Sand	Dumped	-0.3	1.9	1.6
B	Sand	High	-1.3	0.5	-0.8
C	Sand	High	-2.0	0.6	-1.4
D	Clay	Dumped	-0.2	7.8	7.6
E	Clay	High	-1.5	1.1	-0.4
F	Clay	Moderate	-1.2	1.1	-0.1

On the day the 3 m of cover was completed, the pipes with the compacted beddings had not returned to their original diameter.

TIMELAG OF PIPE DEFLECTIONS

Deflection versus time plots are shown on figures 2 through 7. For all test reaches except F, the deflection - time curves are typical. Deflections increase sharply during the first 3 to 6 months following completion of the backfill and then increase at a much slower rate.

Increase in deflection with time is expressed as the "timelag factor," T_f. The timelag factor is defined as the ratio of the deflection measured at some time following completion of the backfill to the deflection measured the day that the backfilling was completed. The timelag factors are summarized in table 8.

Table 8. - Timelag factors

Test reach	Soil type	Degree of compaction	Timelag factors				
			1 month	3 months	1 year	2 years	4 years
A	Sand	Dumped	1.1	1.6	1.7	1.7	1.7
B	Sand	High	1.2	2.0	2.2	2.4	2.4
C	Sand	High	1.4	1.8	2.2	2.2	2.2
D	Clay	Dumped	1.1	1.2	1.5	1.6	1.6
E	Clay	High	1.5	1.8	2.3	2.7	2.9
F	Clay	Moderate	1.7	2.2	4.0	4.3	4.5

The timelag factor for the pipe with the DUMPED bedding material is 1.6 to 1.7. The DUMPED sand bedding, the deflection and time lag have remained essentially the same since the 6-month readings. The largest increase in deflection occurred in the DUMPED clay bedding, where the

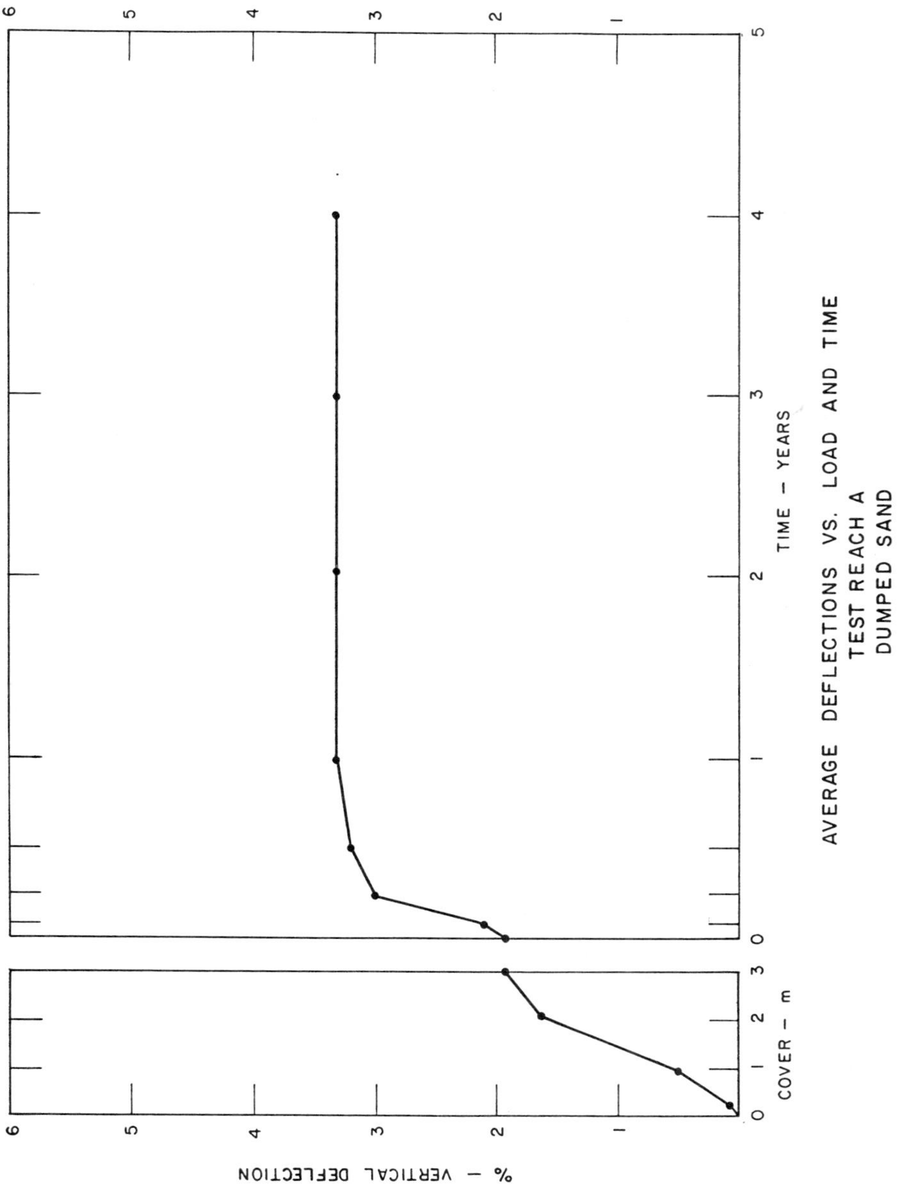

FIG. 2 - - Deflections in Test Reach A

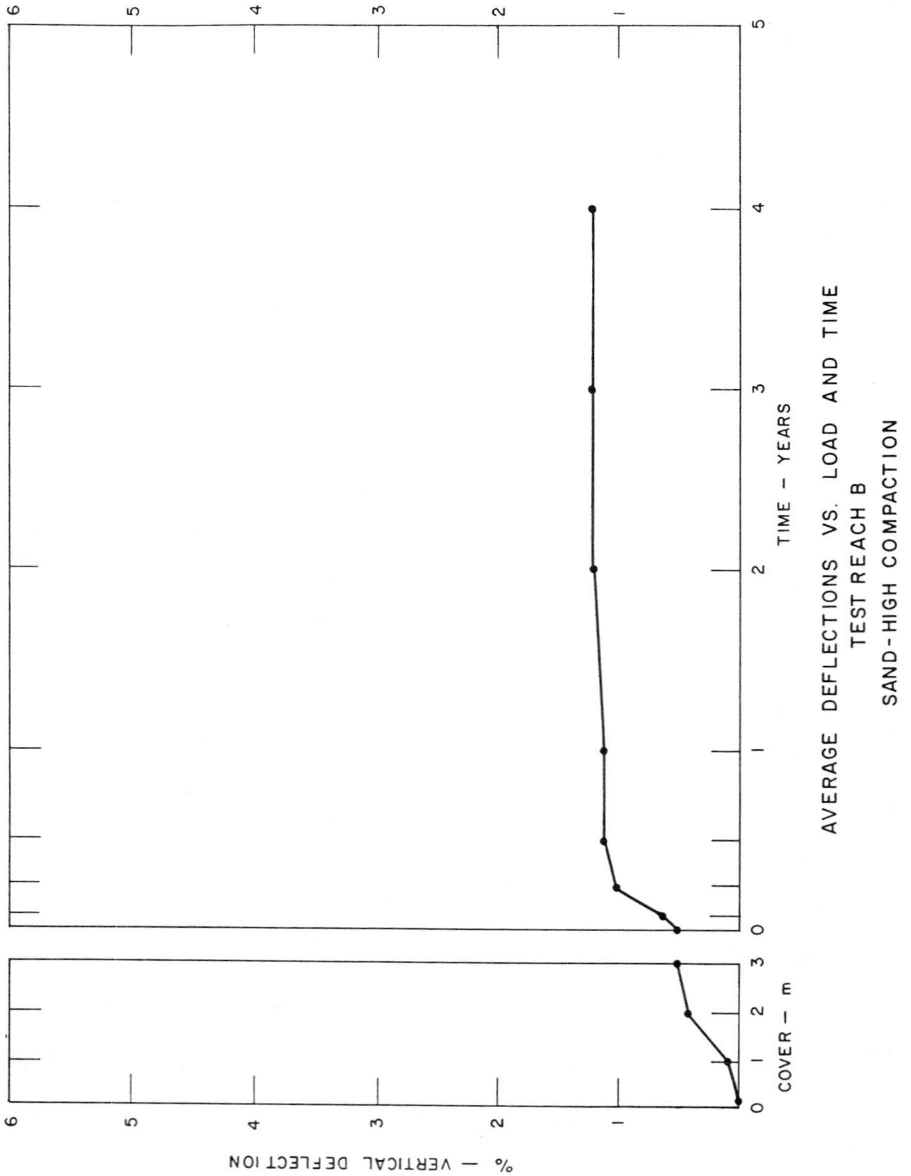

FIG. 3 - - Deflections in Test Reach B

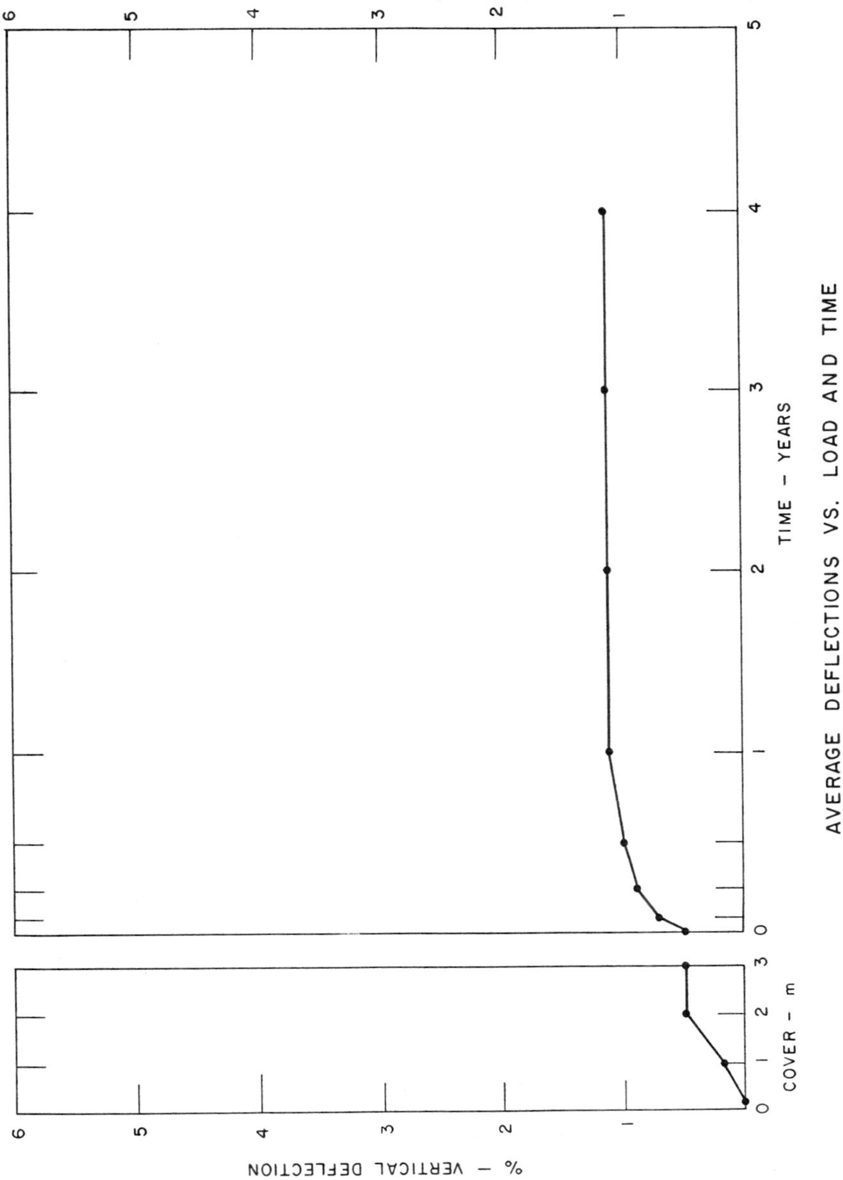

FIG. 4 - - Deflections in Test Reach C

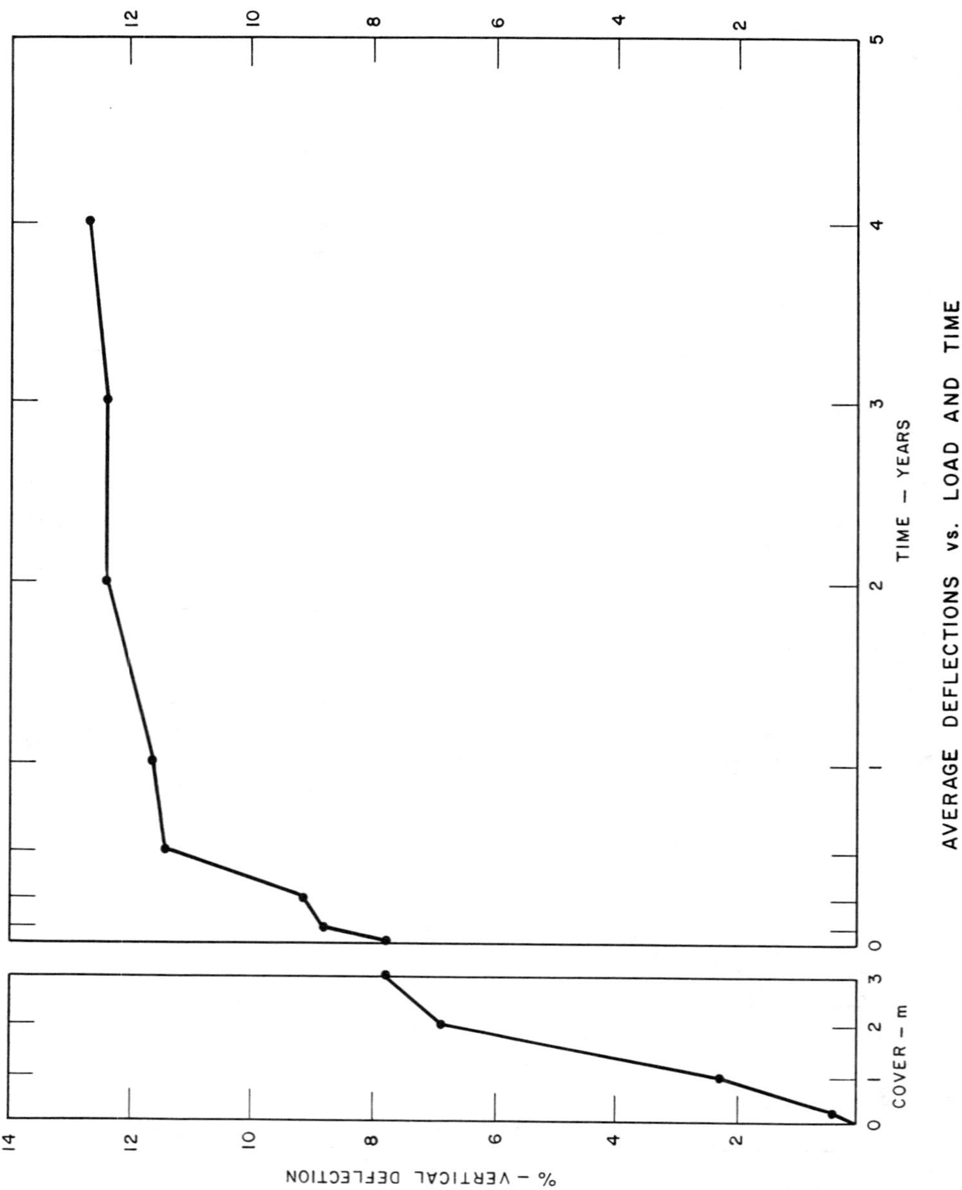

FIG. 5 - - Deflections in Test Reach D

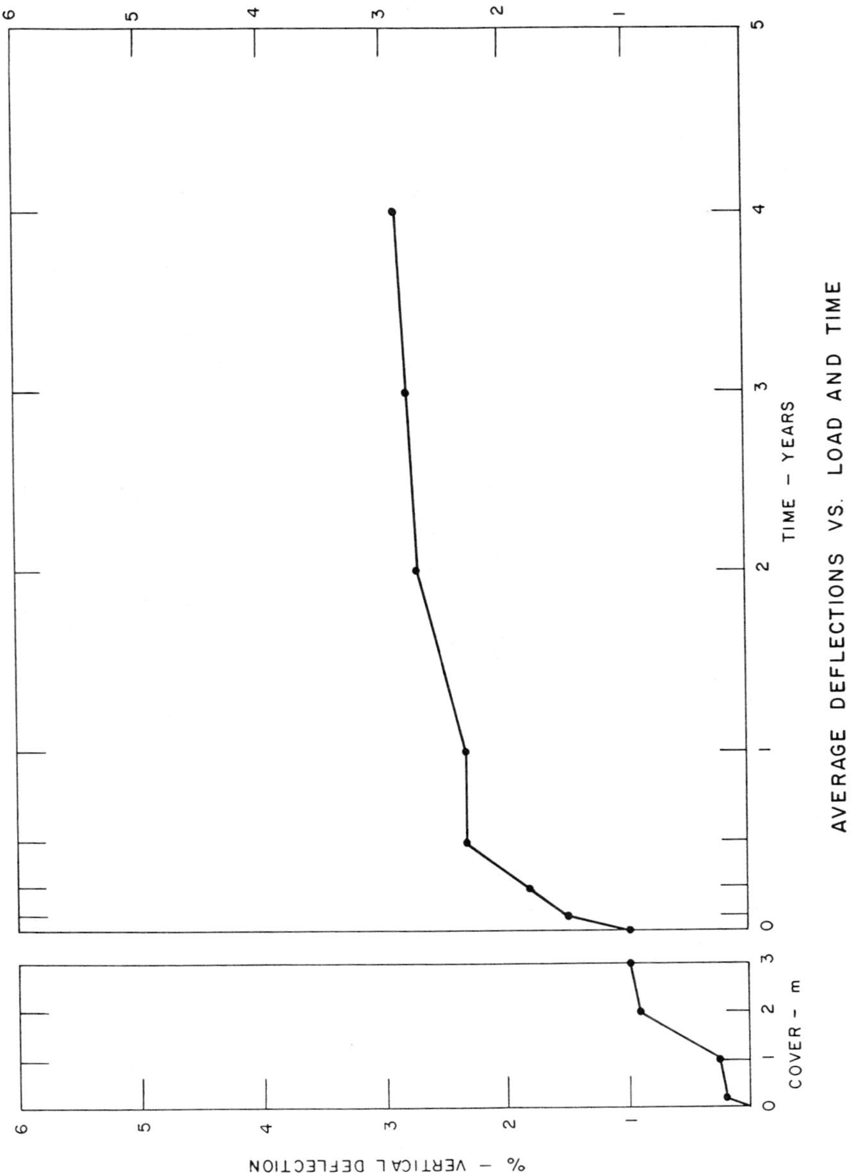

FIG. 6 - - Deflections in Test Reach E

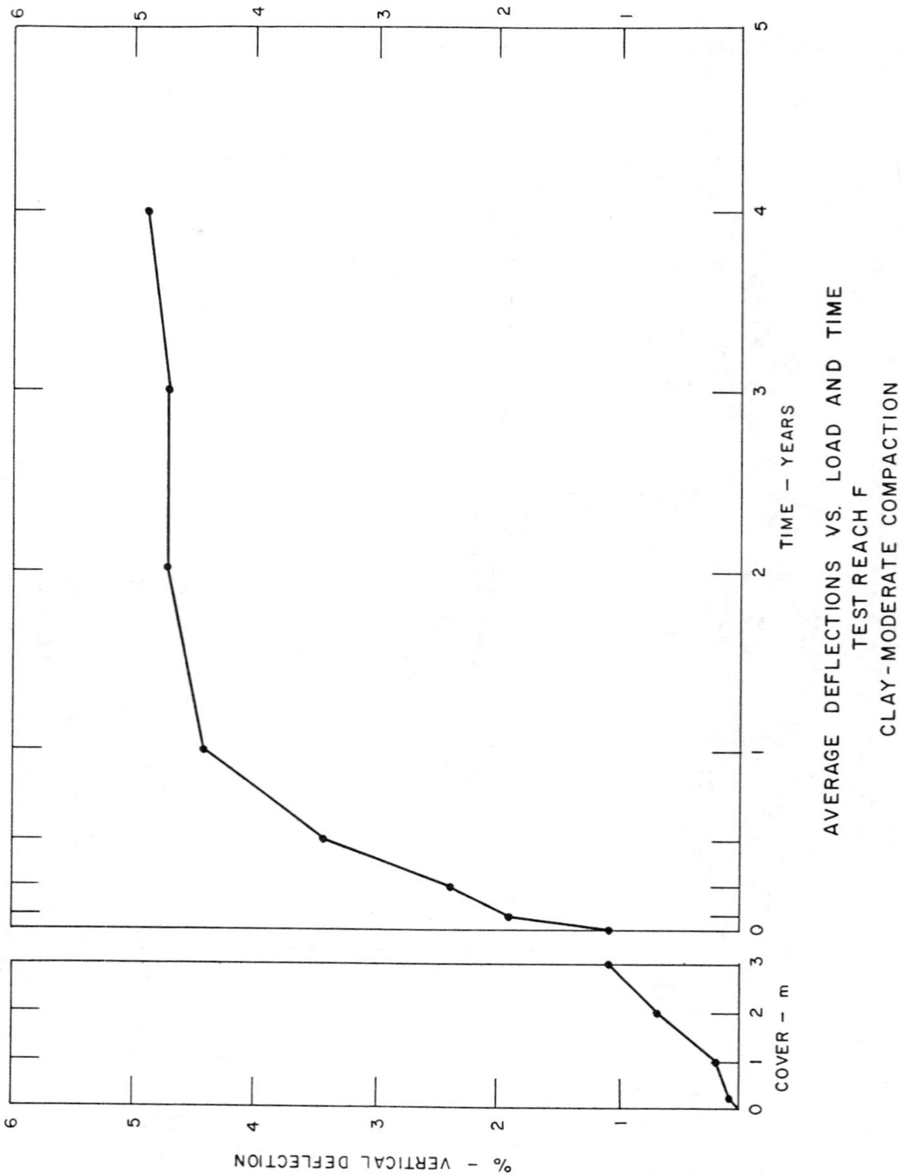

FIG. 7 - - Deflections in Test Reach F

vertical deflection has increased from 11.4 percent at 6 months to 12.7 percent at 4 years; however, the timelag factor only increased from 1.5 to 1.6 during that time.

The timelag factor for the pipe with the HIGH degree of compaction in the bedding material is 2.2 to 2.9 for the 4-year period. For the sand bedding, the increase in the timelag factor for the period from 6 months to 4 years has been small, 2.2 to 2.4 an 2.0 to 2.2. For the clay bedding, the timelag factor has increased from 2.3 at 6 months to 2.9 at 4 years.

The timelag factors for all of the sand beddings have remained constant since the 6-month reading while the pipe installed in the clay beddings have shown continual increases in the time lag factor.

The pipe in the clay bedding with the MODERATE degree of compaction had unexpectedly high timelag factors for the first year, increasing from 1.7 at 1 month to 4.0 at 1 year. However, since the 1-year reading, the timelag factor has leveled off and was 4.5 at the 4-year reading.

Except for test reach F, all the timelag factors are typical. The timelag factors continue to be less for the dumped test reaches than for the compacted test reaches. In the dumped sections, most of the pipe deflection occurs immediately after backfilling because of the poor side support.

The unexpected high increases in deflections in test reach F could be due to either saturation and softening of the clay bedding or excessive load on the pipe due to equipment travel. However, the high water table in this area is continuous for all the test reaches, and equipment travel did not occur.

DEFLECTIONS OF PIPE JOINTS

The joints (bell and spigot) for the test reaches with highly compacted bedding have shown essentially no deflection.

The joints in the two test reaches with dumped bedding (A and D) and in the test reach with the moderate bedding increased in deflection about the same percent as the barrels of the pipe.

The 4-year deflections of the bell and spigot sides of joints in the B and C sections (highly compacted sand) were all 0.1 percent indicating that there was not any significant difference in the deflection of the spigot side of the joint compared to the bell side when the deflection in the barrel of the pipe is about 1 percent.

SUMMARY

Diameter measurements were made over a 4-year period following installation of a 915-mm-inside-diameter RPM pipe as part of a US-USSR joint experiment on buried flexible pipe and are summarized as follows:

1. During installation of the pipe, the maximum increase in vertical diameters occurred at each pipe unit midspan. Just dumping soil in beside the pipe as bedding can oblongate the pipe. In this study, the average oblongation was 0.3 percent for dumped clay and 0.2 for dumped sand.

2. The amount of oblongation is a function of the compactive effort applied to the bedding soil. For the bedding soil compacted to 85 to 95 percent Proctor, the pipe oblongated 1.2 percent. For the bedding soils compacted over 95 percent Proctor, the pipe oblongated 1.3 to 2.0 percent.

3. The pipe with dumped beddings deflected elliptically with the horizontal diameter change about the same as the vertical diameter change. The pipe with compacted beddings deflected rectangularly with the horizontal diameter change about 20 to 60 percent of the vertical diameter change.

4. The joint deflections were between 50 and 100 percent of the pipe barrel deflections in the dumped reaches. In the compacted reaches, the joint deflections were between 0 and 30 percent of the deflection in the pipe barrels.

5. Deflections were greater in the dumped reaches than in the compacted reaches. The 4-year deflections for the reaches with dumped bedding were 3.3 and 12.7 percent while the deflections for the highly compacted beddings were 1.1 to 2.9 percent.

6. Changes in the timelag factors were smaller in the dumped reaches than in the compacted reaches. The 4-year timelag factors for the reaches with dumped bedding were 1.6 to 1.7, while the timelag factors for the highly compacted beddings were 2.2 to 2.9.

7. Test reach F, clay bedding with a moderate degree of compaction, showed the largest increase over the 4-year period with a timelag factor of 4.5. The 4-year deflection was 4.9 percent.

8. The pipe joints in the reaches with highly compacted beddings have shown essentially no deflection.

9. Timelag factors for joints in the test reaches with dumped and moderate bedding were about the same as the timelag factor for deflections in the barrel of the pipe.

10. Timelag factors for all the pipe are the same or only slightly higher at the end of 4 years than the timelag factors at the end of 1 year. The clay beddings with moderate and high degrees of compaction showed the largest increases in timelag factors, 4.0 to 4.5 and 2.3 to 2.9, respectively.

REFERENCES

1. Howard, A. K., Memorandum to Chief, Water Conveyance Branch, Subject: "US-USSR Joint Experiment on Buried Flexible Pipe," Geotechnical Branch Memorandum Reference No. 78-42-72, 79-75, 80-11, 80-29, 81-48, 83-78, and 89-8, Bureau of Reclamation, Denver, Colorado.

Design and Installation

Ian D. Moore[1]

PROFILED HDPE PIPE RESPONSE TO PARALLEL PLATE LOADING

REFERENCE: Moore, I.D., "**Profiled HDPE Pipe Response To Parallel Plate Loading**", Buried Plastic Pipe Technology: 2nd Volume, ASTM STP 1222, Dave Eckstein, Ed., American Society for Testing and Materials, Philadelphia, 1994.

ABSTRACT : The parallel plate load test is used to measure 'pipe stiffness' for HDPE pipe. Pipe stiffness is employed as a measure of pipe resistance to bending deformation as well as a quality control index for the manufacturing process. Unfortunately, the parallel plate test induces a complex state of stress and strain in the pipe, and interpretation of the test results is not straightforward. Simple analysis for a thin circular ring or shell is generally used for these products, but in reality materials like high density polyethylene are viscoelastic (modulus is time and load path dependent) and the depth of the pipe profile may be a significant proportion of the diameter. This paper introduces a three dimensional viscoelastic finite element analysis for HDPE pipe, testing the computational method through comparisons with laboratory data. The analysis is used to examine the nature of pipe response during the parallel plate test. The local distributions of stress and strain through the profile are considered, as well as the effect of loading rate on the pipe response. Conclusions are drawn regarding the ability of conventional thin ring theory to predict circumferential stress and strain, and the implications for pipe design are briefly discussed.

KEY WORDS: finite element analysis, high density polyethylene, pipe, viscoelasticity, local strain, stiffness.

'Pipe stiffness' is generally defined to be the total vertical load applied to a pipe segment, divided by the pipe length and change in vertical diameter. A simple theoretical relationship exists between pipe stiffness PS and the pipe radius r, modulus E and second moment of area I,

$$PS = \frac{EI}{r^3}\left(\frac{\pi}{4} - \frac{2}{\pi}\right)^{-1} \tag{1}$$

This relationship is based on the initial bending stiffness of a thin elastic circular ring responding under two dimensional (plane stress) conditions.

Observations made during HDPE pipe tests clearly confirm that equation (1) is highly idealised. Firstly, HDPE exhibits a viscoelastic (time dependent) response which

[1]Associate Professor, Geotechnical Research Centre, Faculty of Engineering Science, The University of Western Ontario, London, Ontario, N6A 5B9, Canada

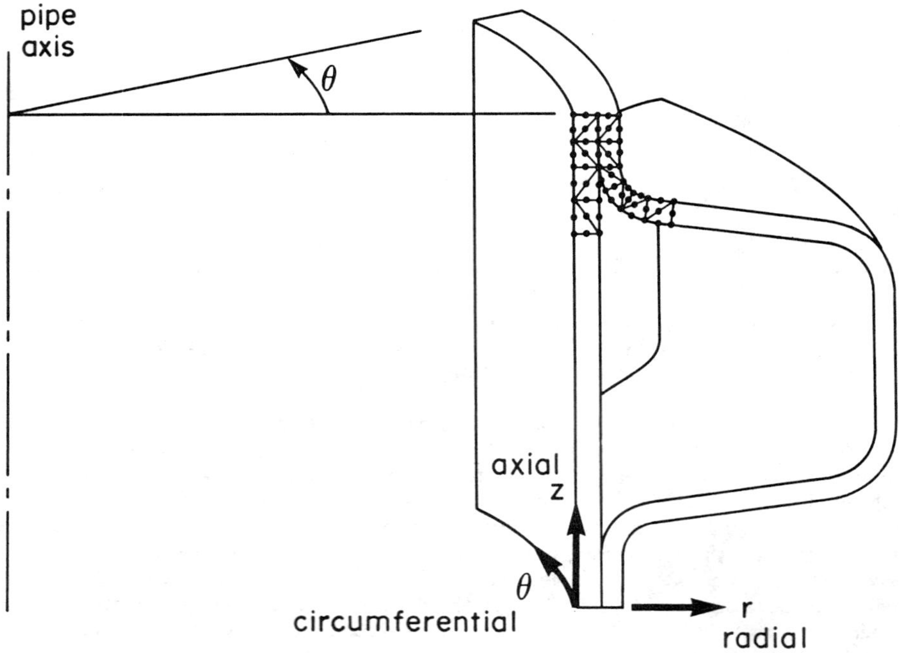

FIG. 1 – Finite element model for three dimensional profiled pipe analysis

makes the apparent value of E a function of the rate of loading and the time after application of loading at which load and deflection are measured. Secondly, many HDPE pipes have profile depth which is a significant proportion of the diameter, so that thin ring theory may be inappropriate. Finally ring deformations may be geometrically nonlinear, that is the ring under two point loading becomes noncircular and this affects the stiffness, [2].

A three dimensional finite element analysis has recently been developed by Moore [4]. This analysis can be used to examine the response of profiled HDPE pipe in the parallel plate test. It can be used to check a number of important design assumptions, both in relation to the parallel plate test and the behaviour of profiled HDPE pipe in the field.

The finite element analysis is described with details provided of the viscoelastic material model used. Elastic predictions are made of local strain distributions for one particular HDPE pipe profile, and these are compared with specific measurements made in the laboratory for HDPE pipe under parallel plate loading. The analysis is then used to make viscoelastic predictions of the load-deflection response of the pipe during a load-unload test, and these are also compared with experimental measurements. Distributions of circumferential and axial stress are predicted and discussed in relation to estimates based on thin ring theory. The effect of loading rate on pipe response is also considered. The paper concludes with a general discussion of the findings.

ANALYSIS OF THE PARALLEL PLATE TEST

The finite element method (e.g. Zienckiewicz, [6]) is an ideal computational procedure for analysing problems in engineering mechanics involving complex geometries and material characteristics. However, the full three dimensional analysis of solids using conventional three dimensional finite element methods is a formidable task since it involves the generation of the three dimensional finite element mesh and the formulation and solution of huge numbers of equations. For corrugated pipes with annular (not helical or spiral) design, use can be made of the axisymmetric geometry to simplify the analysis. A two dimensional finite element mesh is then used to model the geometry and strain fields in the r,z plane, Figure 1, and a Fourier series is used to model variations around the pipe circumference. Pipe response to each Fourier harmonic around the pipe is determined independently, and the full pipe response is assembled using superposition from each of these separate components. The result is a linear three dimensional analysis requiring relatively modest data preparation and computations, Moore [4].

The parallel plate test involves a short length of pipe subjected to load across the vertical pipe diameter, Figure 2a. The actual boundary condition at the crown and invert is one of prescribed displacement, since the pipe rests on a very stiff horizontal surface at the invert and the crown is deformed through a stiff steel plate mounted under the top load platten. Only short segments at the outside of corrugation crests attract load, both at the crown and invert positions.

To simulate the parallel plate test, the short length of annular pipe is modelled and small patches of pressure are applied at the outside of corrugation crests at crown and invert, Figure 2b. In all the analyses reported in this paper, the pipe loading and pipe response is assumed to be symmetric about the horizontal and vertical pipe diameters. The two dimensional finite element mesh used for the HDPE pipe is shown in Figure 3, featuring 1200 six noded linear strain triangular elements. The mesh represents one half of the pipe length 'cut' through the pipe perpendicular to the pipe axis. For the calculations reported here, response was evaluated using one hundred Fourier terms and it was assumed that the plates at crown and springline each made contact over an angle δ measured from the pipe axis of 0.1 radians, Figure 2b. These choices are discussed in detail elsewhere, [3].

CONSTITUTIVE MODEL

Linear viscoelastic finite element analysis is employed, based on conventional rheology utilising sets of 'springs' and 'dashpots', Figure 4a. Unfortunately there is a dearth of test data for HDPE in the literature. However, Chua [1] has reported and used a power law model. This model defines the uniaxial 'secant' relaxation modulus $E = \sigma/\epsilon(t)$ (in psi) at 70°F as

$$E(t) = 7630 + 99507t^{-0.097786} \tag{2}$$

where time is expressed in minutes, Figure 4b. Undertaking a simple curve-fitting exercise for a multi-Kelvin model with one independent spring and nine Kelvin elements in series, the independent spring is found to have modulus 1000 MPa. The Kelvin springs have moduli 3580MPa, 0.85x3580MPa, 0.85^2x3580MPa, ..., 0.85^8x3580MPa. The Kelvin dashpots have viscosities 0.503MPa.days, 5.03MPa.days, 50.3MPa.days, up to 5.03x10^7MPa.days. This conventional rheological model is used to predict the time dependent response of the pipe.

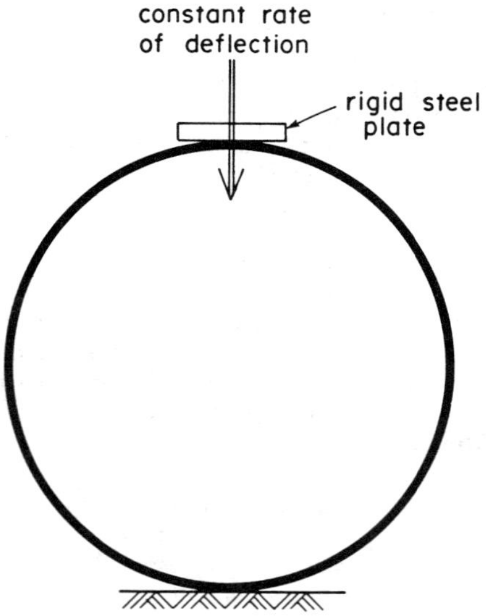

a. Loading conditions for the parallel plate test

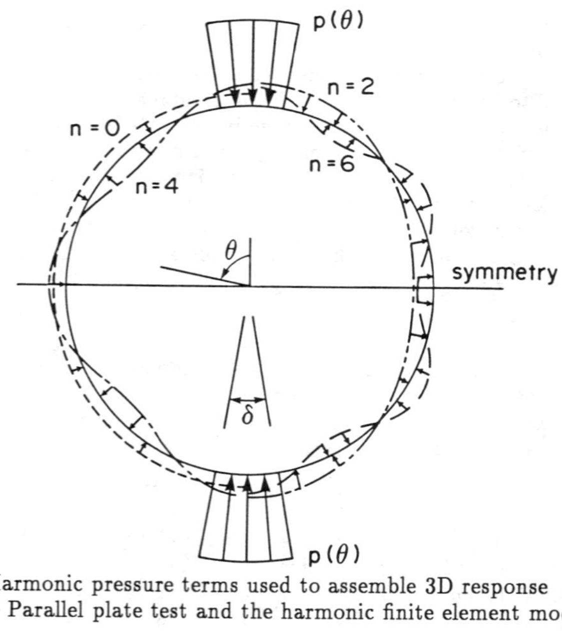

b. Harmonic pressure terms used to assemble 3D response

FIG. 2 – Parallel plate test and the harmonic finite element model

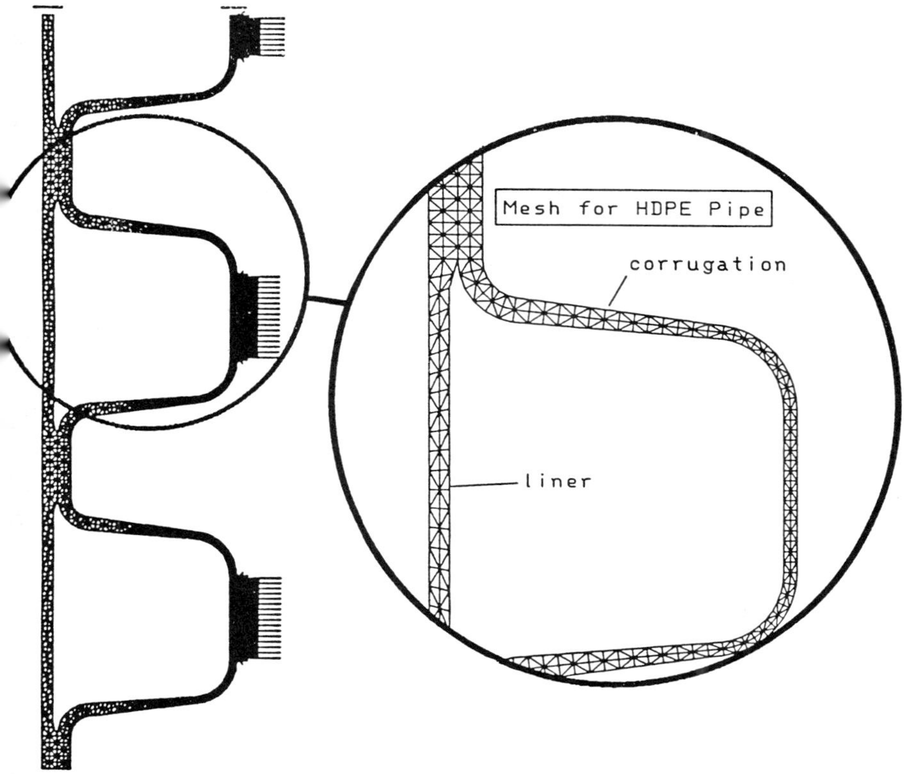

FIG. 3 – Finite element mesh for analysis of the HDPE pipe test

TABLE 1 –
omparison of predicted and measured values of circumferential strain for corrugated HDPE pipe

location on the corrugation (1)	predictions using FEM (2)	measurements		
		near pipe center side 1 (3)	near pipe center side 2 (4)	side 1 (5)
outside crest	5100$\mu\epsilon$	B:5900$\mu\epsilon$	J:5300$\mu\epsilon$	D:5600$\mu\epsilon$
sloping sides	2500$\mu\epsilon$	A:3000$\mu\epsilon$	C:3000$\mu\epsilon$	
inside trough	-4200$\mu\epsilon$	G:-4300$\mu\epsilon$		
inside liner	-2200$\mu\epsilon$	H:-2100$\mu\epsilon$	I:-4000$\mu\epsilon$	E:-1600$\mu\epsilon$
near junction	-4000$\mu\epsilon$		F:-4200$\mu\epsilon$	

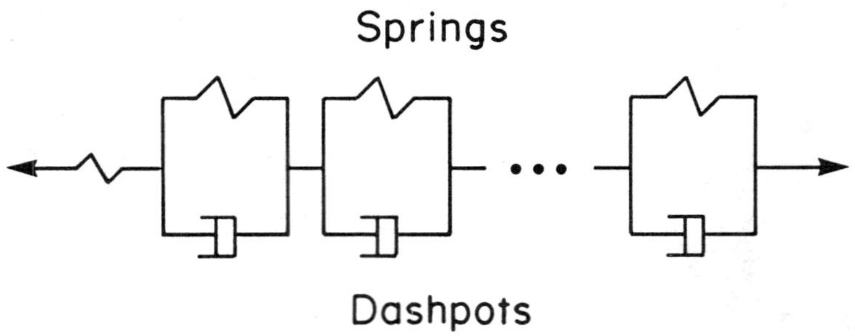

a. Spring and dashpot model for HDPE

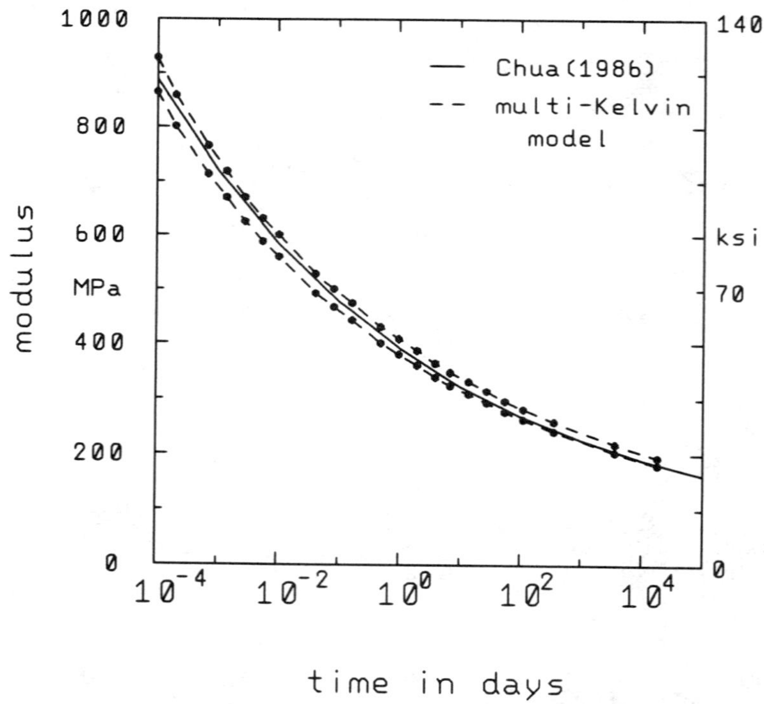

b. 'Creep' modulus from the Chua model and the multi-Kelvin model
FIG. 4 – Viscoelastic models for HDPE

STRAIN DISTRIBUTIONS

Laboratory tests have been performed on the HDPE pipe to measure local strains under parallel plate loading. A 450mm diameter pipe of length 320mm was instrumented with a series of resistance strain gauges, Moore [3]. Figure 5 shows the gauge locations, labelled 'side 1' and 'side 2'. Gauge locations *A* to *J* were instrumented with planar or stacked rosettes. The pipe was loaded so that vertical pipe diameter decreased steadily by 12mm per minute until a total diameter change of 24mm. After holding this position for 1 minute, the pipe was unloaded at the same deformation rate. Local strain measurements were taken throughout this period, in addition to a record of the total vertical load acting across the pipe.

Figure 6a and Figure 6b show the distribution of circumferential and axial strain estimated for the springline of the pipe at vertical pipe diameter change of 24mm (all quantities shown are in microstrain, tension positive). Uniform elastic properties were assumed for the HDPE, with Poisson's ratio 0.46. Young modulus does not need to be specified explicity for this elastic calculation - the strains were simply scaled to correspond to the vertical pipe diameter change of 24mm. Boundary conditions for the finite element analysis feature one pipe end restrained against deflection and rotation (the pipe centreline) while the other end is left completely free. Uniform radial pressures were applied at both corrugation crests.

Various regions of tensile and compressive strain can be identified in Figures 6a and 6b. The circumferential strain distribution is similar to that which would be calculated using thin ring theory. Tensile strains of up to 6100$\mu\epsilon$ are noted in the corrugation at the 'fibres' most distant from the pipe axis. Compressive strains develop in the liner. There is some local bending apparent in the liner and the corrugation crest (strain contours at those locations are not consistently parallel with the neutral axis). For axial strains, the response is more complex. Local bending produces strain gradients in the liner and the crest of the corrugation - the former has largely tensile strains and the latter is generally in compression. The largest tensile strain (>4000$\mu\epsilon$) occurs at the junction of the liner and corrugation, and the largest compressive strain (<-4000$\mu\epsilon$) is at the outside of the corrugation crest.

Tables 1 and 2 show measurements of circumferential and axial strain respectively together with estimates of local strain at the strain gauge locations. In general, it appears that an elastic analysis of the pipe based on uniform modulus across the pipe section and around the pipe circumference leads to reasonable estimates of the circumferential strain quantities. The value of circumferential strain for gauge *I* appears to be significantly different to those measured at locations *E* and *H* but apart from this one value, the measured and estimated strains in the circumferential direction are generally within 15%.

Estimates of axial strain do not appear to be quite as good. A noticeable discrepancy occurs for gauge location *G*, as well as at the centreline of the liner. At the liner centreline the difference between the average measurement and the finite element prediction is 900$\mu\epsilon$. The measurement at *G* has no equivalent measurement for comparison, so it is unclear whether that one reading is spurious, or whether there really is a substantial error in the theoretical prediction. At other locations the theoretical predictions of axial strain are within 500$\mu\epsilon$ or 20% of the average of the measured values.

TABLE 2 –
Comparison of predicted and measured values of axial strain for corrugated HDPE pipe

location on the corrugation (1)	predictions using FEM (2)	measurements		
		near pipe center side 1 (3)	near pipe center side 2 (4)	side 1 (5)
outside crest	-4800$\mu\epsilon$	B:-4000$\mu\epsilon$	J:-4100$\mu\epsilon$	D:-4300$\mu\epsilon$
sloping sides	-800$\mu\epsilon$	A:-700$\mu\epsilon$	C:-800$\mu\epsilon$	
inside trough	2200$\mu\epsilon$	G: 5600$\mu\epsilon$		
inside liner	-1000$\mu\epsilon$	H:-100$\mu\epsilon$	I:-300$\mu\epsilon$	E:100$\mu\epsilon$
near junction	2500$\mu\epsilon$		F:2600$\mu\epsilon$	

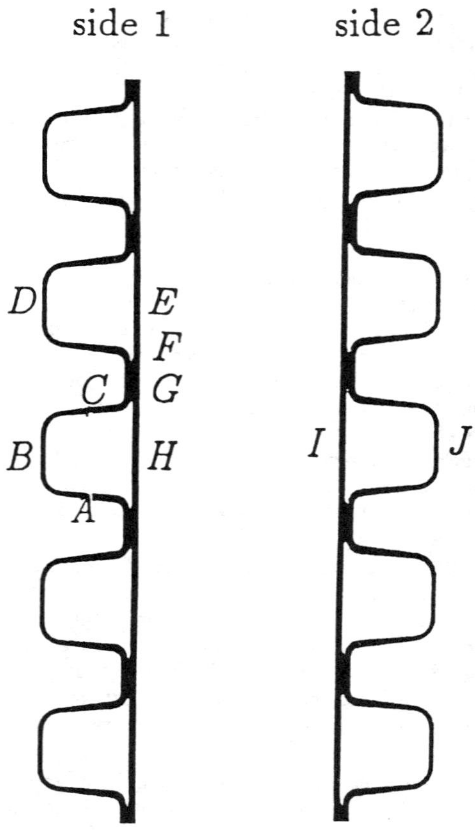

FIG. 5 – Location of the strain gauges at the springline of the pipe specimen

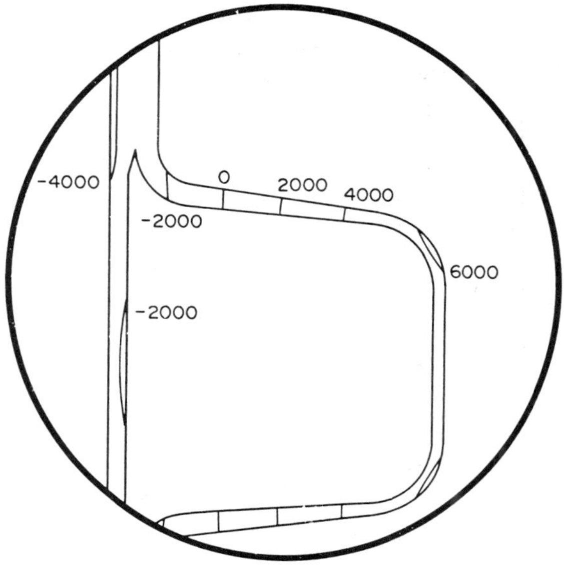

a. Circumferential strain distribution

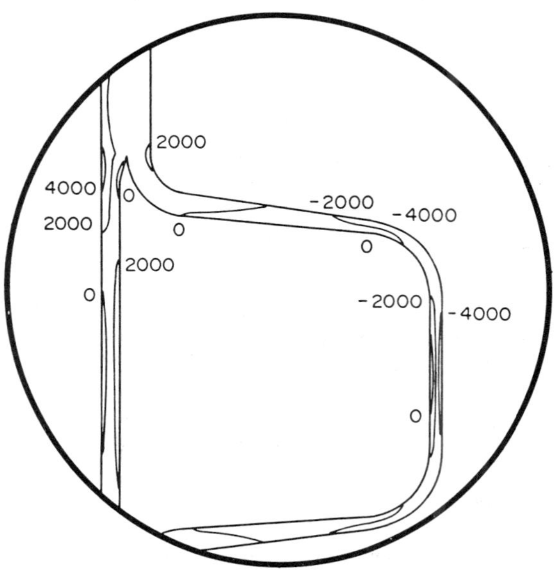

b. Axial strain distribution

FIG. 6 – Finite element estimates of local strain at the HDPE pipe springline; at 24mm vertical diameter change

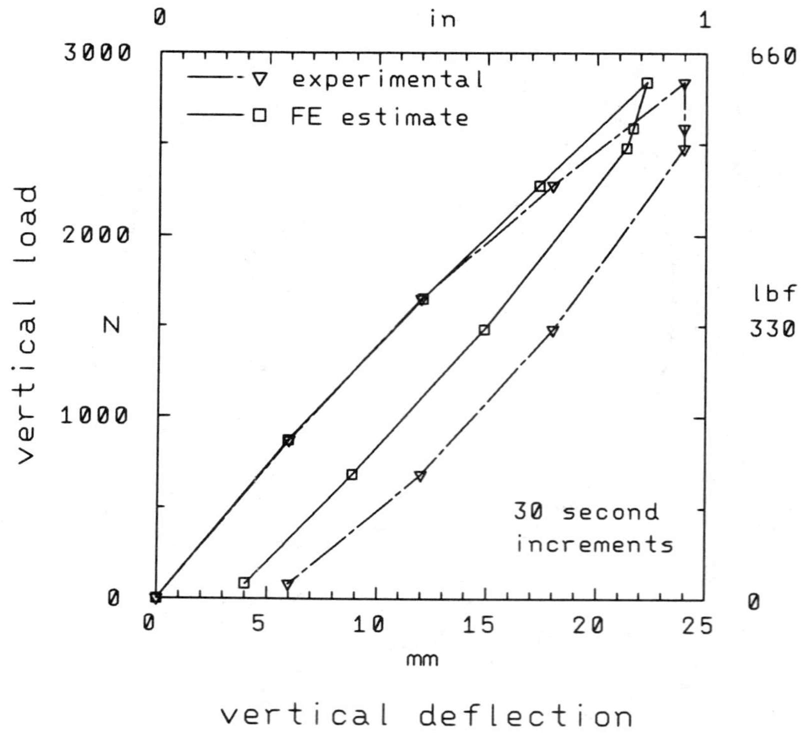

FIG. 7 – Experimental measurements and finite element predictions of the load deflection response of the HDPE pipe

VISCOELASTIC PIPE RESPONSE

Linear viscoelastic finite element analysis can be employed to estimate the load deflection response during the test and to estimate the relationship between applied load and strain at any particular location.

The experimental load deflection curve and the finite element estimate are shown in Figure 7. The analysis is very close for most of the initial loading stage, but near the peak load the deflection predictions drop to 3 mm less than the measured values. The unloading response is similar in shape to that measured, but remains at reduced deflection.

Figure 8 shows a finite element estimate of the strain versus load response together with experimental measurements at gauges *D* and *J*. This comparison reveals that the three dimensional finite element analysis with viscoelastic properties reported by Chua [1] provides very reasonable estimates of strain versus load for this parallel plate test. At the peak load, the discrepancy of $600\mu\epsilon$ is similar to that reported earlier in Table 1.

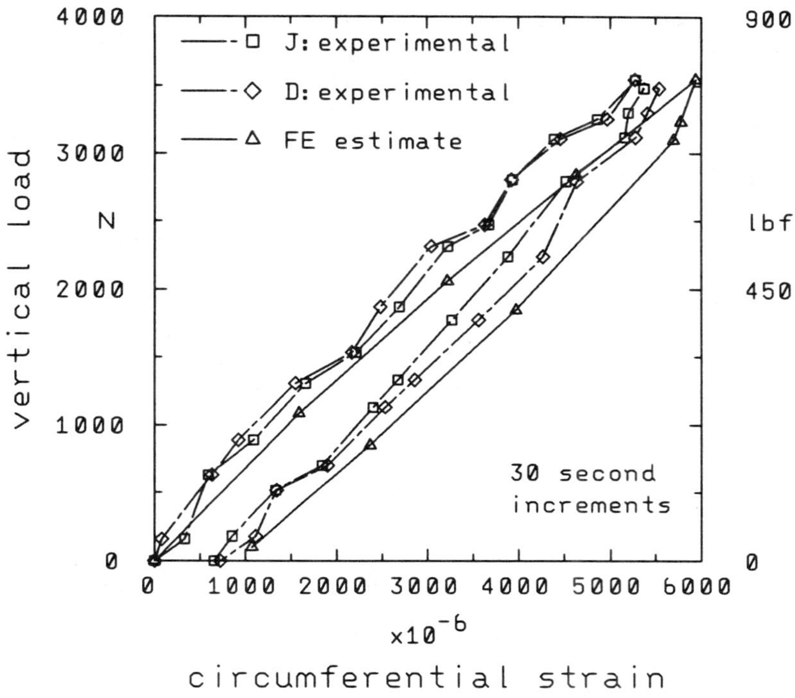

FIG. 8 – Experimental measurements and finite element predictions of the local strain response of the HDPE pipe

STRESS DISTRIBUTIONS

With a theoretical model capable of successfully predicting pipe response during the parallel plate load test, it is straightforward to investigate a number of issues. For example, the stresses that develop through the pipe profile are of interest. Those stresses will be greatest at the point in time where load is highest. Figures 9a and 9b, show distributions of circumferential and axial stress at the springline of the pipe at that peak load.

Firstly, the circumferential stress distribution is similar in pattern to the circumferential strain distribution examined earlier. Peak tensile stress is about 9MPa, occurring at the inside of the flat section of the corrugation crest, as well as at the outside of the curved sections of the profile. This tensile stress for the 450mm diameter pipe at 5% vertical deflection, is 40% of the peak stress for HDPE under uniaxial tension (peak or ultimate stress is about 22MPa, although thus figure depends on temperature and loading rate). Peak compressive circumferential stress is somewhat over 6MPa, but is of no great concern given the superior performance of HDPE in compression.

Tensile stresses in the axial direction reach a maximum of close to 6MPa at the inside of the corrugation near the 'extreme fibre'. The local bending in this segment leads to compressive stresses almost equal and opposite at the outside surface. Local bending in the section of the liner which spans the corrugation also produces stresses of equal and opposite magnitude at internal and external surfaces, but these have magnitude 3MPa. While axial stresses do develop at the springline of this pipe during the parallel plate load test, they are less in magnitude than the stresses that develop in the circumferential direction.

EFFECTIVENESS OF THIN RING THEORY

With theoretical estimates and experimental measurements available for local strain and/or stress in the profiled HDPE pipe, it is possible to evaluate the effectiveness of conventional thin ring theory for estimating strain and stress in this type of pipe product.

Firstly, 'thin ring theory' reveals that during the parallel plate test the bending moment M_{sp} at the springline of the pipe and vertical pipe deformation ΔD_v are given by [5]

$$M_{sp} = Wr(\frac{1}{\pi} - \frac{1}{2}) \tag{3}$$

$$\Delta D_v = \frac{2Wr^3}{EI}(\frac{1}{\pi} - \frac{1}{4}) \tag{4}$$

for applied load W. For the pipe at peak load W=2840N, integration of the circumferential stresses estimated during the finite element analysis yields a circumferential moment of 119N.m. The value calculated from equation (2) is 124.N.m, just 4% higher. Using the same applied load, the radius to the neutral axis of the profile calculated to be 241mm and the second moment of area about the neutral axis for the whole pipe length calculated as 4.180x10^{5}mm^4, standard circumferential stress σ_θ calculations can be made based on thrust $N = -W/2$, moment M_{sp} and distance d to the extreme fibre:

$$\sigma_\theta = N/A + \frac{M_{sp}d}{I} \tag{5}$$

At the corrugation crest the distance from the neutral axis is 30mm, implying a 8.2MPa tension, and the distance to the inside surface of the liner is 12.4mm resulting in 3.0MPa compression. The estimate of tension is very close to the values shown on Figure 9a, but the ring theory estimate of maximum compression is half of the three dimensional estimate. Significant stress redistribution appears to be occurring at the junction of the liner and the corrugation which is increasing compressions in this region. Local bending in the liner also appears to be generating compressions larger than expected.

Estimates of local strain can also be made using thin ring theory: the hoop strain N/EA combines with bending strain associated with changes in curvature M/EI. These estimates require the use of an equivalent 'elastic modulus'. From equation (4), E is estimated to be 1180MPa. This modulus permits calculation of hoop strain from cross-sectional area 2100mm^2, viz. -0.057%. Change in curvature is 0.000252. At a distance 30mm above the neutral axis circumferential strain is expected to be 7000$\mu\epsilon$. At the inner fibre, the circumferential strain should be -3700$\mu\epsilon$. These are similar to those reported in Table 1. The tensile value is in excess of the finite element estimate as well as the measured strains. The compressive value is slightly smaller in magnitude.

In summary, it appears that the use of conventional two dimensional ring theory will provide reasonable estimates of bending moment and circumferential stress and strain provided the applied loads are known. Estimates of circumferential stress and strain on the

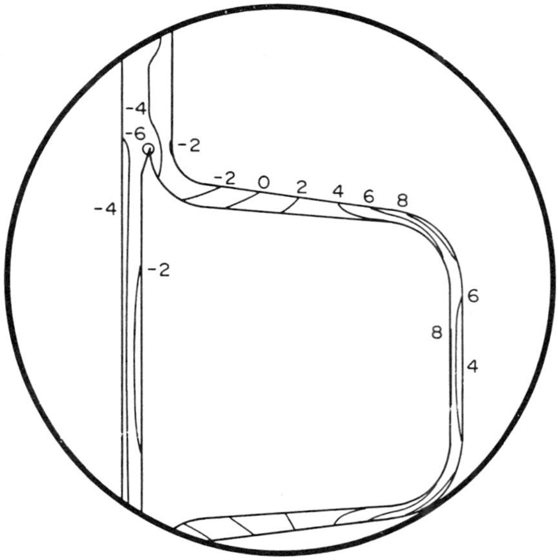

a. Circumferential stress in MPa

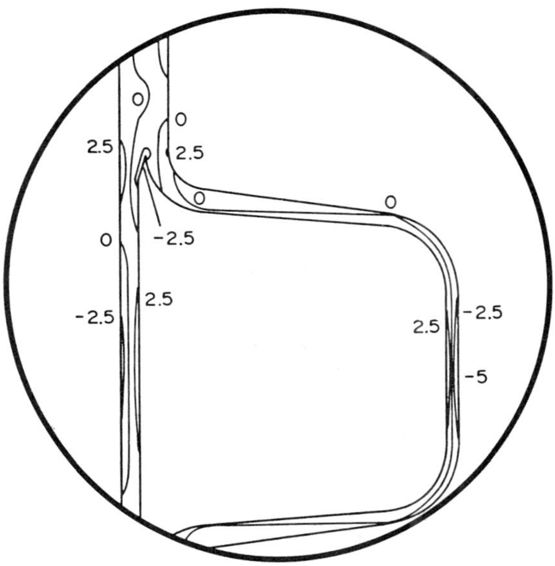

b. Axial stress in MPa

FIG. 9 – Finite element estimates of local stresses at the HDPE pipe springline

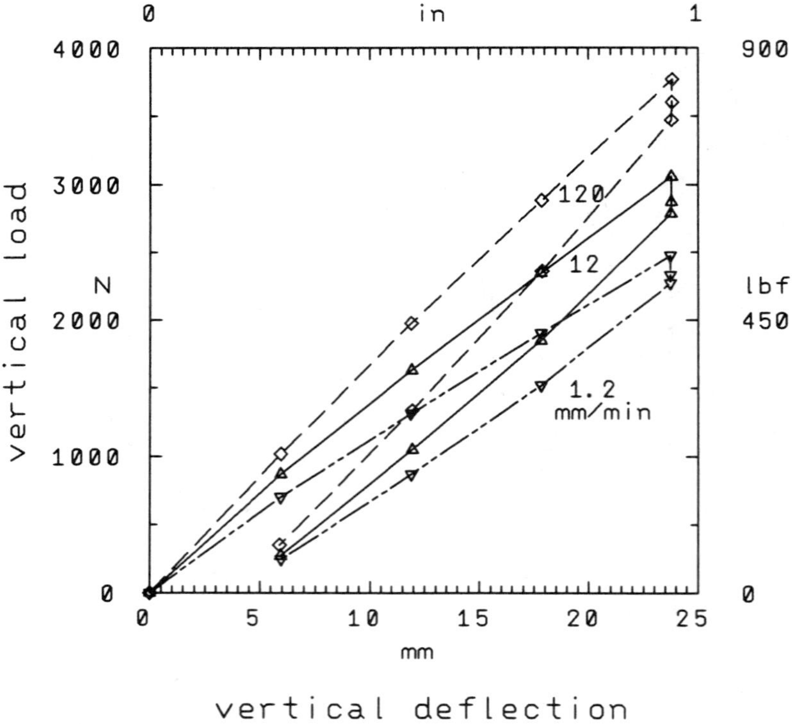

FIG. 10 – Finite element estimates of load deflection response for HDPE pipe at three different loading rates

corrugation of the pipe at the location most distant from the pipe axis are either close or somewhat conservative. Estimates of circumferential stress on the pipe liner are lower than those from three dimensional theory. Naturally, two dimensional thin ring theory cannot provide estimates of axial stress or strain. These conclusions should apply to pipes in the field as well as those under parallel plate loading.

EFFECT OF LOADING RATE

It has been recognised for some time that rate of loading during the parallel plate test is an important issue for HDPE pipes. The ASTM standard D2412 requires a steady decrease in vertical pipe diameter of 12mm (0.5in) per minute and this was the deformation rate used during the experimental work on the HDPE pipe. Using the finite element analysis, loading rates of 120mm/min, 12mm/min and 1.2mm/min are specified to examine the effect on pipe stiffness. Figure 10 shows three estimates of HDPE pipe load-deflection based on the three dimensional viscoelastic finite element analysis. These reveal that by increasing the loading rate tenfold to 120mm/min, the measured load at 24mm vertical diameter change (and therefore the pipe stiffness) increases by 12%. When loading rate is decreased tenfold to 1.2mm/min, the load at 24mm deflection decreases by 19%. The changes in rate of loading must be substantial before noticeable changes in loads and therefore pipe stiffness occur. This has been confirmed by tests in the laboratory where rate of loading was increased twofold; stiffness values changed only slightly.

In addition to the pipe loading configurations already considered, the finite element model can simulate other loading patterns. These will be the subject of future studies, to determine the nature of buried HDPE pipe response to live and dead load.

CONCLUSION

Three dimensional finite element analysis has been used to estimate the behaviour of local surface strains for segments of HDPE pipe subjected to short term parallel plate loading. Experimental measurements of surface strain at the pipe springline have been used to examine the performance of that computer model. Elastic predictions of circumferential and axial strain were generally found to be quite reasonable - within five and fifteen percent of the measured values.

The power law model of Chua [1] was used to develop rheology to permit viscoelastic predictions of the HDPE pipe response. These predictions were generally successful, although more work is needed to improve deformation predictions during pipe unloading. Estimates of strain versus load compared well with experimental results. It appears that the analysis provides good predictions of pipe stiffness, in addition to the three dimensional distributions of strain.

The distributions of circumferential and axial stresses were estimated. For the pipe profile considered, the circumferential tensions that developed at the springline were higher than any axial tensions. Thin ring theory was shown to provide reasonable or somewhat conservative estimates of circumferential tensions at the springline, but to underpredict circumferential compression in the liner. It was found that loading rate during the parallel plate test must be changed substantially before measured load and pipe stiffness is much affected. Further work is needed to examine the effectiveness of the analysis for predictions of HDPE pipe response over longer periods of time, and much slower loading rates. The three dimensional analysis should also be used to examine the response of buried HDPE pipe and to examine the implications of local bending observed at certain points in the pipe profile.

ACKNOWLEDGEMENTS

Support for the experimental work described in this paper has been provided by Big 'O' Inc. and Dow Canada as well as the National Research Council of Canada through the IRAP program. The theoretical analysis was developed with the assistance of research and equipment grants to the author from the Natural Sciences and Engineering Research Council of Canada. J. Chirico and W. Logan performed the laboratory tests and their contributions are gratefully acknowledged.

REFERENCES

[1] K.M. Chua. *Time-dependent interaction of soil and flexible pipe.* PhD thesis, Texas A & M University, 1986.

[2] H.B. Harrison. Force measurement with proving rings. Technical Report R302, School of Civil Engineering, The University of Sydney, 1977.

[3] Ian D. Moore. Local strain in corrugated pipe : experimental measurements to test a numerical model. *Journal of Testing and Evaluation, ASTM*, (to appear), 1993.

[4] Ian D. Moore. Three dimensional time dependent models for buried hdpe pipe. To appear in *The Proceedings of the Eighth International Conference on Computer Methods and Advances in Geomechanics*, H.J. Siriwardane, editor, Morgantown, WV, USA, May 1994. A. A. Balkema.

[5] O.C. Young and J.J. Trott. *Buried rigid pipes : Structural design of pipelines.* Elsevier Applied Science, 1984.

[6] O.C. Zienkiewicz. *The finite element method in engineering science.* McGraw-Hill, 1979.

Amster K. Howard[1]

INSTALLATION OF PLASTIC PIPE USING SOIL-CEMENT SLURRY

REFERENCE: Howard, Amster K., "**Installation of Plastic Pipe Using Soil-Cement Slurry,**" Buried Plastic Pipe Technology: 2nd Volume, ASTM STP 1222, Dave Eckstein, Ed., American Society for Testing and Materials, Philadelphia, 1994.

ABSTRACT: Soil-cement slurry used in buried pipe installations has become an increasing popular choice for contractors. Flexible pipe, including PVC and RPM, as well as rigid pipe are being installed using this technique. The ingredients of the soil-cement can vary, but typically is a combination of soil, portland cement, and water. In most cases, the pipe trench is trimmed to a semicircular shape that is only slightly larger than the pipe diameter. The soil-cement is used to fill the gap between the pipe and the in situ soil. Accordingly, the native trench material must be able to provide adequate supporting strength to the pipe. The consistency of the soil-cement can vary from a fluid (slurry) to a mixture with a 25 cm slump depending on the placement requirements.

The consistency, ingredients, and placement dimensions can all vary as long as two basic requirements are met:

1. The material must be placed so that there is complete contact between the pipe and the in situ soil.

2. The unconfined compressive strength of the hardened material is at least 700 kN/m^2 (100 lb/in^2) at 7 days.

The most suitable soil to use is a silty sand with the fines content not exceeding about 30 percent. This allows native soils from the trench excavation or from nearby the construction site to be used. Cementitious fly ash has been used in place of cement and bentonite has been added to improve pumping characteristics.

KEY WORDS: soil-cement, slurry, plastic pipe, construction, soils, pipelines, soil treatment, casper, testing

---

Soil-cement slurry was first used by the Bureau of Reclamation as an alternate method of pipe installation in 1963. As an option in Reclamation specifications, soil-cement slurry was used infrequently

[1] Research Civil Engineer, U.S. Bureau of Reclamation, PO Box 25007, Denver CO 80225.

until the mid-1980's. As the cost of compacted embedment and labor rose, the use of soil-cement became increasingly more attractive to contractors. Reclamation personnel, faced with doing more with fewer people, liked the minimum monitoring required with soil-cement slurry. Some projects requested that soil-cement slurry be the only method allowed in the specifications. Concurrently, similar materials (variously known as controlled low-strength material or CLSM, soil-cement grout, flowable fill, unshrinkable fill, flowable mortar, etc.) are increasingly being used for filling voids and abandoned tanks, trench backfill, and foundation backfill in tight locations. This renewed interest in soil-cement slurry has required evolving placement techniques and specification requirements. This paper documents Reclamation experiences with this technique and discusses the current requirements.

As shown on figure 1, the standard method for the Bureau of Reclamation of installing pipe is to excavate a flat bottom trench with either vertical or sloping walls with the width of the trench bottom equal to the pipe diameter plus 45 cm. Compacted earth is placed to 0.7 of the outside diameter of flexible pipe. The soil required is a clean, free-draining cohesionless material that typically has to be processed and imported to the site. Reclamation requires that this select material be compacted to at least 70 percent relative density as defined in ASTM D 4253 "Standard Test Method for Maximum Index Density and Unit Weight of Soils Using a Vibratory Table."

Reclamation specifications allow for contractors to use an alternate method, soil-cement slurry, for installing pipe 30 cm in diameter and larger when the native soil is firm enough to provide the necessary support for the pipe. In this alternate method, the trench is trimmed to a semicircular cross section, as shown on figure 2, that is slightly larger than the pipe. The soil-cement slurry is used to fill the gap between the pipe and the in situ soil. The only purpose of the soil-cement slurry is to fill this gap so that the load from the pipe is transferred to the in situ soil. The in situ soil must be able to provide the same supporting strength to the pipe as the compacted select material used in the standard method of pipe installation.

The consistency and ingredients of the soil-cement can vary, but typically is a combination of soil, portland cement, and enough water so that the mixture has the consistency of a thick liquid. In this form, the slurry flows readily into openings and provides a hardened material that has a strength greater than the untreated soil used in the mix and greater than the adjacent in situ material.

As discussed later, there have been many variations in the materials, consistency, and trench configurations. Any mixture or consistency can be used as long as two basic requirements are met, as follows:

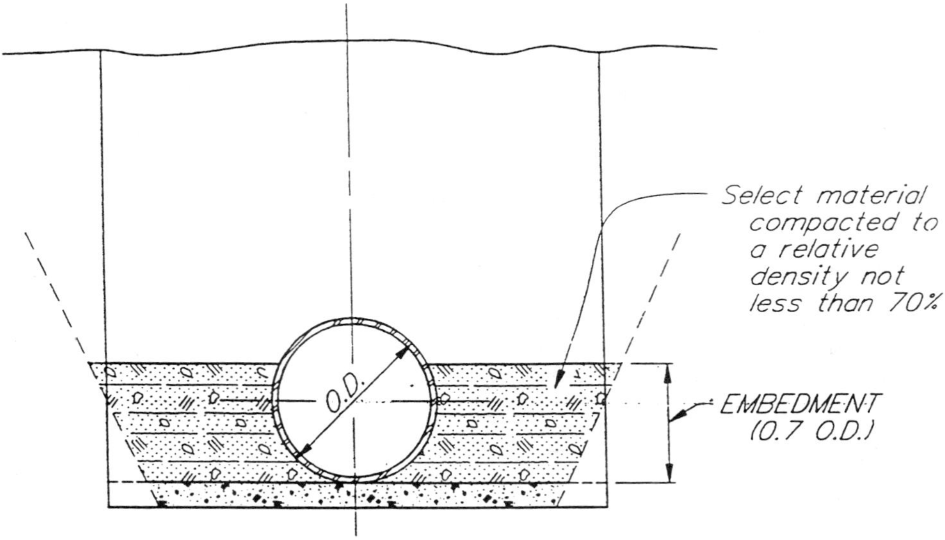

Figure 1. - Standard USBR Pipe Installation

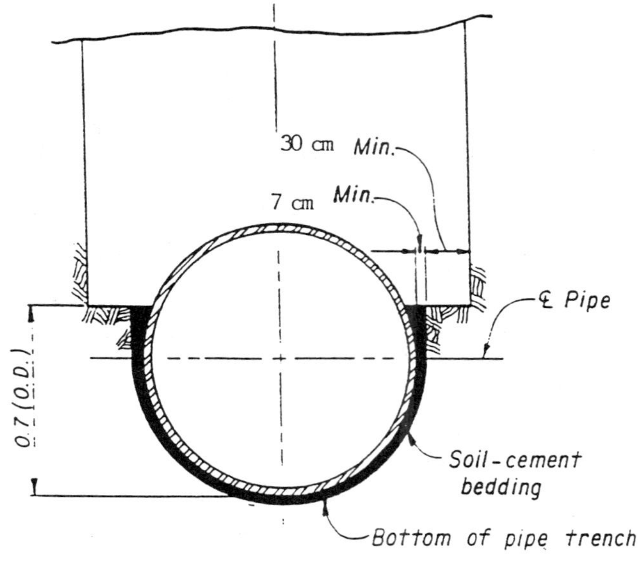

Figure 2. - Soil-Cement Slurry for Pipe Installation

1. The material must be placed so that there is complete contact between the pipe and the in situ soil.

2. The unconfined compressive strength of the hardened material is at least 700 kN/m^2 (100 lb/in^2), but not more than 1,400 kN/m^2 (200 lb/in^2) at 7 days.

Trial mixes are usually tested before construction to determine the adequacy of the ingredients and the mixture.

CONSTRUCTION

The excavation, mixing, and placement of the soil-cement slurry can be accomplished in many different ways depending on the size and length of the pipeline, the characteristics of the in situ soil, and the availability of materials.

Excavation

The portion of the trench above the upper surface of the soil-cement slurry is excavated so that at the top of the soil-cement there is a minimum of 30 cm (1 foot) between the pipe and the trench wall to allow working space beside the pipe. The two excavation methods most commonly used are (1) cut a flat bottom trapezoidal trench down to the upper surface of the soil-cement slurry so that a trencher or other device can operate in the bottom of the trench to cut out the circular shape for the soil-cement, or (2) use a backhoe with a semicircular bucket to excavate the upper portion of the trench and then excavate the semicircular lower portion of the trench.

The trimmed lower portion of the trench for the soil-cement slurry would ideally be a semicircular shape with the diameter 15 cm (6 inches) larger than the outside of the pipe (about 7 cm on each side for placement of the slurry). This requires a minimum amount of excavation and material handling. In addition, a minimum amount of soil-cement slurry is needed to fill the annular space between the pipe and the in situ soil. This semicircular area has been excavated with a trenching machine with a special arrangement of the cutting teeth or with a backhoe with a semicircular bucket. Both must be especially adapted for the size pipe to be used. If the pipeline is short, or there are several different diameter pipes, this may not be economical. Other shapes that have been used are shown on figure 3. A smooth, closely controlled section is not always possible. Materials such as shale, siltstone, claystone, or sandstone may only be able to be excavated with the material removed as chunks so that a neat trimmed surface is not possible. Some materials may have to be blasted, leaving an irregular trench shape. In these cases, more volume of soil-cement slurry will be required but its use may still be economically justified.

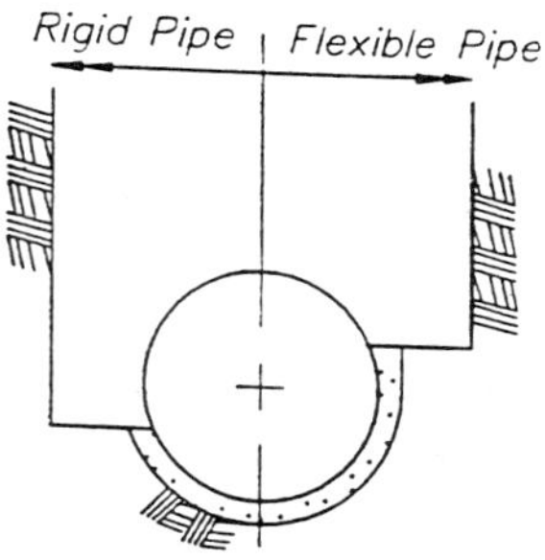

TYPE A

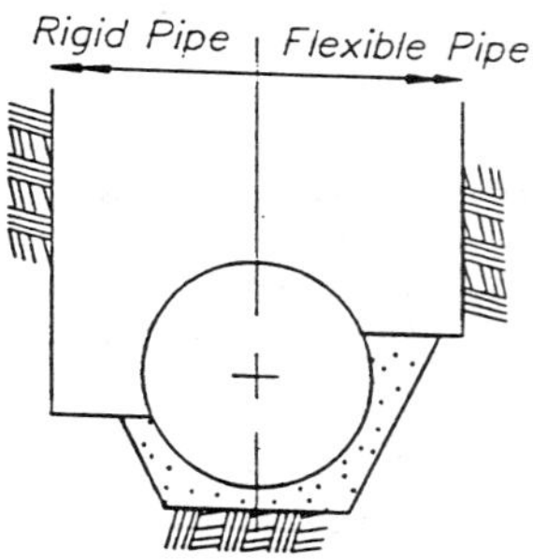

TYPE B

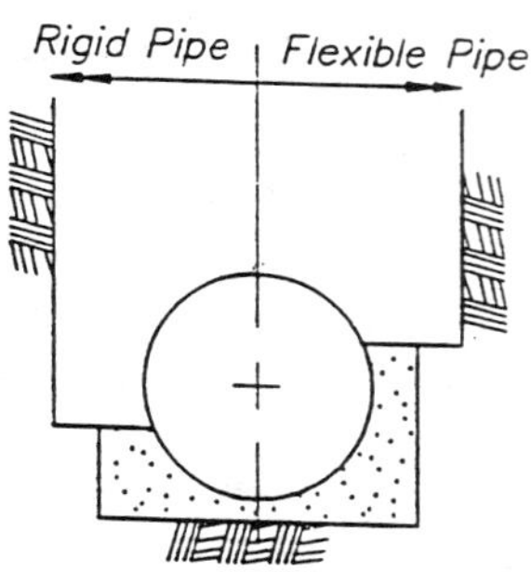

TYPE C

Figure 3

Acceptable Trench Shapes

Mixing

Any method of mixing the materials can be used as long as the soil-cement slurry has a uniform consistency and appearance just before placement.

If material excavated from the trench is used in the soil-cement slurry, a trench-side traveling batch plant or a portable batch plant close to the site can be used.

If a town is nearby, most contractors elect to obtain the soil-cement slurry from a commercial ready-mix plant. Either conventional concrete sand can be used or special bins of silty sand can be incorporated in the batching operation.

Placement

The pipe is laid on two soil pads leveled to the proper grade. The soil pads can be of any material that will be "weaker" or of less stiffness than the hardened soil cement. This is necessary to avoid having two hard spots beneath the pipe which may result in concentrated point loads. While the soil may be loose or compacted, pipeline grade is easier to maintain with compacted soil than with uncompacted soil. Sand bags have been used effectively. The soil-cement slurry is placed on one side of the pipe and allowed to flow under the pipe until it can seen on the other side. Then the soil-cement slurry is added to both sides until the space between the pipe and the in situ soil is filled. Rodding or vibration may be needed to keep the soil particles in suspension so the material flows easily. Soil-cement slurry made with clean concrete sand tends to bleed excessively and vibration is normally required to move the mixture into all openings. Soils containing 15 to 30 percent fines can be normally placed without vibration or rodding. When stiffer mixes are required [20- to 25-cm (8- to 10-inch) slump], vibration is required. Because of the difficulty of placing the soil pads (or sand bags) on very steep slopes, the pipe is often laid directly on the trench bottom. When there is not a space beneath the pipe for the soil-cement slurry to flow from one side to the other, visual inspection must be used to ascertain that the soil-cement slurry is filling all spaces around the pipe.

Soil-cement slurry has the potential to float the pipe, particularly with light-weight plastic pipe. To prevent this, the soil-cement may need to be placed in two stages. The first placement should have reached initial set before the remainder is placed. Backfill should not be placed over the soil-cement slurry until the soil-cement has reached its initial set. Small cracking of the soil-cement surface is not a problem, but deep cracking and shrinkage must be prevented. Since moisture is beneficial to curing, a 15-cm (6-inch) cover of moist earth should be placed over the soil-cement slurry if the backfill is not to be placed over the soil-cement within 8 hours after placement.

Cement will not hydrate (harden) below about 5 °C (42 °F), so the temperature of the soil-cement slurry should be maintained well above that temperature. Accordingly, several specification provisions discuss temperature. Soil-cement should not be placed when the air temperature is below 4 °C (40 °F) unless the temperature is 2 °C (35 °F) or above and rising. The temperature of the soil-cement slurry must be 10 °C (50 °F) or greater at the time of placement. The soil-cement slurry must be protected from freezing. Before the initial set, an insulation blanket should be used and an earth cover can be used after the initial set. If the air temperature is 10 °C (50 °F) or less, then the moist earth cover over the exposed soil-cement slurry should be at least 45 cm (18 inches) thick. The soil-cement slurry is not to be placed in pipe trenches when the trench bottom or walls are frozen or contain frozen material.

MATERIALS

Any materials and mix design can be used as long as two basic requirements are met:

1. The mixture is of a consistency such that it completely fills the space between the pipe and the in situ soil; and

2. The hardened mixture has an unconfined compressive strength of at least 700 kN/m^2 (100 lb/in^2), but not more than 1,400 kN/m^2 (200 lb/in^2) at 7 days (1 week).

Soil-cement slurry is typically a combination of soil, portland cement, and water. However, cementitious fly ash may be used in place of the cement, pozzolans may be added, and the consistency may range from a fluid to a high-slump material. If the soil-cement slurry is to be pumped, bentonite may be added to improve the flow characteristics of the slurry through the delivery hose.

Soils

The most suitable soil to use is a silty sand with the fines content not exceeding about 30 percent (30%). This allows native soils from the trench excavation or from the area to be used. The fines must be nonplastic or have a low plasticity.

Clean concrete sands have often been used when the soil-cement slurry is obtained from local ready-mix plants. Bleeding can occur with clean concrete sands and vibration is often necessary to keep the mixture in suspension as it is placed in the space between the pipe and the soil and to ensure that it flows completely around the pipe.

The presence of fines can prevent the bleeding, result in higher compressive strengths, and reduce the need for vibration.

The allowable maximum particle size is related to the dimension of the annular space between the pipe and the in situ soil to prevent bridging as the mixture is poured into the space. The maximum particle size in the soil should not exceed one-eighth of the open distance between the pipe and the trench wall or 38 mm (1-1/2 inches), whichever is less. For the typical 7-cm (3-inch) gap, a maximum particle of 19 mm (3/8 inch) should not be exceeded. However, with a larger gap, the maximum particle size can be increased as long as the larger soil particles stay in suspension.

The soil used must not contain a quantity of organic impurities that would affect the time of set and strength. The presence of organic impurities should be checked in accordance with ASTM C 40 "Standard Test Method for Organic Impurities in Fine Aggregate for Concrete."

Clay balls in the soils do not become penetrated by cement, and they become points of weakness in the hardened mixture. The maximum percentage of clay balls, on a wet mass basis, should not exceed 10 percent (10%). The maximum size of any clay balls should not exceed 15 mm (1/2 inch) to prevent bridging of materials in small areas.

Cement

The cement used should meet the requirements for cement used in concrete as described in typical structural concrete specifications. Pozzolans, particularly fly ash, may be added to reduce the cement content to save costs. The cement content will typically be 5 to 10 percent (5 to 10%) by dry mass (about one to three sacks per cubic yard) of the soil, in order to obtain the 700 kN/m^2 (100 lb/in^2) strength requirement.

Cementitious fly ashes (type C) may be used in place of cement as long as the strength requirement is met. Fly ash sources can be extremely variable and close monitoring of the properties of the fly ash may be required.

Water

The water used should meet the requirements for water used in concrete.

Consistency

In most cases, the mixture has the consistency of a thick liquid so that it flows readily into openings and fills any voids. The water-cement ratio should not exceed 3.5. Typically, the water-cement ratio is between 2 and 3.

On slopes, or when a single pipe is individually installed using a trench shield, a fluid consistency is not appropriate and the mixture should be placed with a high slump [20 to 25 cm (8 to 10 inches)] to prevent the soil-cement slurry from flowing down the trench. With a stiffer mix, bleeding and keeping all the soil particles in suspension is not a problem, although vibration will probably be required to work the material into complete contact with the pipe and the soil.

INSPECTION AND TESTING

One of the reasons that soil-cement slurry for pipe installation is popular with Reclamation field personnel and is attractive for contractors is the reduced amount of inspection and testing and the virtual elimination of reworking and retesting compacted soil. A consistent mix and placement are much easier to maintain with soil-cement slurry than with the placement and compaction of soil. Soil requires recompaction when the field density tests performed on the soil result in unacceptable low densities. Typical pipeline construction can typically result in 10 to 25 percent of the soil having to be recompacted, especially at the beginning of construction when experimentation with the appropriate compaction procedures is necessary to obtain the required density.

Testing

Testing involves checking the mix ingredients for specifications requirements and determining the compressive strength of the mixture.

A minimum of two 15- by 30-cm (6- by 12-inch) compressive strength cylinders should be prepared to represent each sampled batch. In the initial stages, preparation of three cylinders is recommended in order to obtain representative data.

The preparation and testing of the soil-cement slurry cylinders are in accordance with ASTM D 4832-88. "Standard Test Method for Preparation and Testing of Soil Cement Slurry Test Cylinders." The compressive strength is to be determined at least once for every 150 linear meters (500 linear feet) of pipe placed, or at least twice per shift. The testing frequency should be increased in the initial stages and maintained until routine construction procedures and consistent test results are established. The compressive strength tests are performed to check the adequacy and uniformity of the mixes.

The soil-cement must have sufficient compressive strength to transfer the load directly from the pipe to the in situ soil. The minimum value of 700 kN/m^2 (100 lb/in^2) at 7 days for compressive strength has a built-in safety factor since results are not known for 7 days. If strengths start falling below the minimum, the soil-cement is still adequate but changes to the mix must be made so the strengths start to meet the minimum. The compressive strength values should not be any higher than 1,400 kN/m^2 (200 lb/in^2). Higher strengths serve no purpose and mean an unnecessary amount of cement is being used. A low-strength material would also be easier to remove if the pipe had to be repaired or replaced.

The gradation of the material should be checked for percent fines and maximum particle size. In addition, checks on the presence of clay balls are required.

ADVANTAGES AND DISADVANTAGES

The advantages of using soil-cement slurry over a compacted soil bedding would be:

1. Reduction in volume of excavated material and associated earth moving costs.

2. Time, personnel, and equipment required to vibrate soil-cement slurry is less than that required to vibrate select material. Often, the consistency and mix ingredients of the soil-cement slurry will not require vibration.

3. Time and equipment required to test compressive strength cylinders of soil-cement are less than that required to perform field and laboratory density tests for quality control.

4. Simplicity and consistency of soil-cement slurry placement virtually ensures proper installation initially, thus eliminating recompacting and retesting involved with compacted soil.

5. Soil-cement can be made using native silty sands while the select material required for pipelines generally has to be processed and imported.

6. Use of a select material creates a french drain around the pipeline that may require construction of trench plugs in areas where natural groundwater movement would be affected by the french drain. Soil-cement slurry around the pipe would eliminate this concern.

The disadvantages of using soil-cement slurry over a compacted soil would be:

1. The reduced space in the trench prohibits the checking of gaskets at bell-and-spigot joints.

2. Test cylinder compressive strength results are not available until 7 days after installation. Any problems with the mix design may delay the placement of backfill over the pipe.

CONCLUSIONS

Soil-cement slurry has been used by the Bureau of Reclamation for pipeline installation since 1963. The use of soil-cement slurry has become more popular in the last decade due to the rising cost of placing compacted embedment around buried pipe. The consistency and ingredients of the soil-cement slurry can vary, but typically it is a combination of soil, portland cement, and water. In most cases, the pipe trench is trimmed so that the space is only slightly larger than the pipe diameter. The soil-cement slurry is used to fill the gap between the pipe and the in situ soil. Thus, the native trench material must be able to provide adequate supporting strength to the pipe.

The consistency, ingredients, placement dimensions can all vary as long as two basic requirements are met:

1. The material must be placed so that there is complete contact between the pipe and the in situ soil.

2. The unconfined compressive strength of the hardened material is at least 700 kN/m^2 (100 lb/in^2) and not more than 1,400 kN/m^2 (200 lb/in^2) at 7 days (1 week).

REFERENCES

[1] Lowitz C. A., and DeGroot G., "Soil-Cement Pipe Bedding, Canadian River Aqueduct," Journal of the Construction Division, ASCE, vol. 94, No. CO1, January 1968.

[2] Day R., "Cradling Method Speeds Large-Diameter Pipeline," Excavating Engineer, March 1965 (Canadian River Project).

[3] "Grout Bedding Speeds Pipe Laying" Western Construction, November 1968 (Westlands Water District).

[4] "Cement-Treated Pipeline Bedding" Publication No. PA0011.01, Portland Cement Association, Skokie, Illinois, no date.

Larry J. Petroff[1]

DESIGN METHODOLOGY FOR HIGH DENSITY POLYETHYLENE MANHOLES

REFERENCE: Petroff, L.J., "Design Methodology for High Density Polyethylene Manholes," Buried Plastic Pipe Technology: 2nd Volume, ASTM STP 1222, Dave Eckstein, Ed., American Society for Testing and Materials, Philadelphia, 1994.

ABSTRACT: Polyethylene manholes are used in piping systems for sewage, storm water run-off, landfill leachate, and industrial effluent. Many manholes are used in applications no more than a few feet deep, where loading is minimal and almost any vertical structure will work. However, for manholes of any significant depth and for manholes located below the groundwater level, substantial earth and groundwater loads may exist. Although often ignored, these loads can approach critical values for the manhole. Evaluation of the manhole, embedment and loading is a crucial task of the design engineer. This paper presents a design methodology that may be used for such applications.

The design methodology considers both the ring-directed and the axial-directed effects of the applied earth pressure. Ring loads subject the manhole wall to compressive strains which, if not limited, can cause buckling. These loadings are usually intensified by groundwater. Axial strains occur in the manhole wall due to the downdrag of the surrounding soil. Proper design limits axial strains and provides an adequate safety factor against buckling.

KEY WORDS: flexible pipe, buried pipe, manholes, thermoplastic, stiffness, strain, buckling

High Density Polyethylene (HDPE) has many applications in sanitary landfills and other corrosive environments. Leachate collection pipes, storm water run-off collection systems, gas gathering pipes, dual containment pipes, collection structures, linings, and manholes are made from HDPE. HDPE manhole structures have been developed for placement directly in landfill cells (with a small amount of surrounding granular material). HDPE manholes are in service in sanitary landfills in Germany with depths approaching 100 meters. Loading on these structures may be very high and special designs are usually required to relieve the effects of fill settlement.

Less ambitious but equally critical applications occur in the U.S. where HDPE manholes have been installed to depths approaching 30 meters in fill materials. The design of these manholes is critical. Yet, for even relatively shallow manholes (3 to 5 meters) such as those encountered on municipal sewerage applications, soil settlement may exert considerable downdrag force. As a result, shallow manholes, as well as, deep manholes, must be properly designed to withstand the anticipated loads.

[1]Supervisor of Engineering, Technical Department, PLEXCO®, 1050 Busse Highway, Suite 200, Bensenville, Illinois 60106

While a large amount of research work along with codes and standards writing has been done for plastic pipes, little has been done for plastic manholes. Hossain and Lytton have published the results of a computer model study on HDPE manholes [1]. The study used a proven finite element program and incorporated the viscoelastic properties of HDPE. Field verification of the model study is in progress. Until this is completed, the designer of HDPE manholes must rely on semi-empirical, linear-elastic approaches for developing a design methodology. The methodology given herein is one such approach. It is believed to be conservative in that it ignores the ability of HDPE manholes to undergo stress reduction by complying with the fill material.

PERFORMANCE LIMITS

Manhole shafts are fabricated from one or more lengths of pipe, joined by either fusion welding or bell and spigot. Since the shafts are constructed from flexible pipe, they must be checked for the same performance limits as buried pipe, which includes ring deflection, hoop stress (or strain), and ring buckling. Like pipes, radially directed loads acting on a manhole cause ring deformation and ring bending strain. Unlike pipes, the radial load varies along the length of the manhole. See Fig. 1.

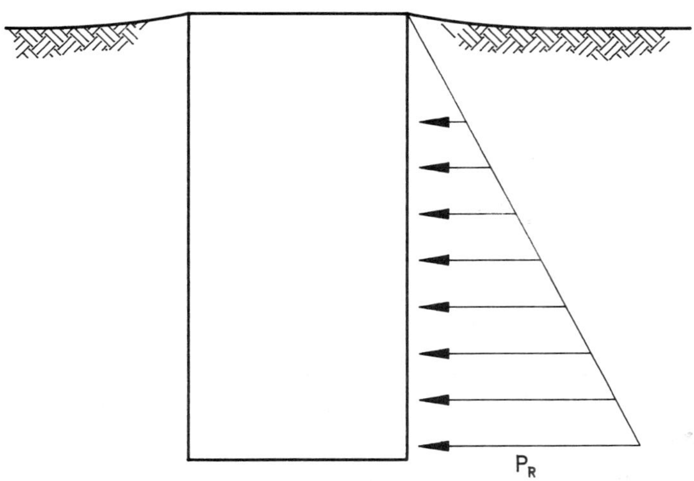

Figure 1: Radial Pressure Acting on Manhole (Assumed distribution for design)

In addition to radial stresses, considerable axial stress may exist in the manhole wall as a result of "downdrag". Downdrag occurs as the backfill soil surrounding the manhole consolidates and settles. Axial load is induced through the frictional resistance of the manhole to the backfill settlement. Therefore, the manhole must also be checked for axial compressive strain and axial buckling. See Fig. 2.

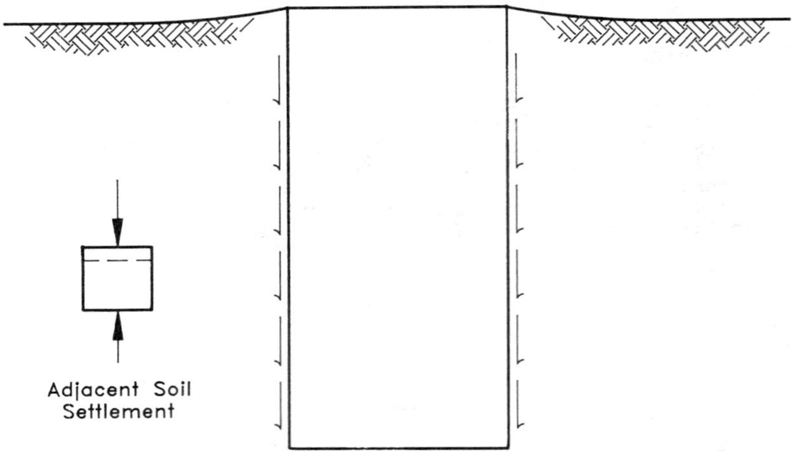

Figure 2: Downdrag Force Acting on Manhole (assumed for design)

EARTH LOADING ON MANHOLE SHAFTS

The earth load on a manhole shaft depends in part on how the shaft is installed. For instance, if the manhole is placed in a predrilled hole, the maximum horizontal earth pressure acting on the manhole would be less than the pressure occurring in undistributed soil. When the hole is drilled, the earth is free to move laterally into the hole and develop active pressure as well as radial arching. For manholes installed in embankments, disposal sites, or other fills, the method of placement of the fill around the manhole will affect the resulting earth pressure. If the fill is placed in compacted lifts around the manhole, the soil may be prestressed horizontally and exert a pressure that exceeds the active earth pressure and approaches the at-rest earth pressure.

The flexibility and material characteristics of the manhole also affect the loading. For instance, manholes made from HDPE pipe experience a reduction in radial earth pressure with time as the manhole undergoes circumferential (ring) compression as a result of stress relaxation.

RADIAL PRESSURE

Gartung, Prühs, and Hoch reported on the design of landfill manholes at the Second International Landfill Symposium [2]. They are associated with the LGA-Ground Construction Institute in Nuremburg, Germany. LGA regulates landfills in southern Germany where landfill manholes have been constructed from concrete, thermoset plastic, and HDPE. In their report, they recommended using the active earth pressure (for practical design purposes) based on measurements taken by Prühs and on the ability of the material placed around the manhole to accept tangential stresses and thus relieve some of the lateral pressure.

The stress relaxation properties of HDPE allow it to shed load into surrounding soil. Thus, HDPE manholes undergo stress reduction after installation. This allows mobilization of horizontal arching and the active earth pressure can be assumed for design purposes. If the active earth pressure is modified to take into account uneven compaction around the perimeter of the pipe, the radially-directed design pressure is given by Eq. 1. [3]

$$P_R = K_A \gamma H \left[2-\sqrt[3]{\cos^2\alpha + 0.5 \sin^2\alpha}\right] \quad (1)$$

Where: α = Angle of orientation in horizontal plane, deg
P_R = Applied Radial Pressure, kPa
γ = Soil Unit Weight, kN/m^3
H = Height of Fill, m
K_A = Active Earth Pressure Coefficient

$$K_A = \tan^2\left(45-\frac{\phi}{2}\right) \quad (2)$$

Where: ϕ = Angle of Internal Friction, deg

The expression used for uneven compaction has a maximum value of 1.21 at α equals 300°. So, Eq. 1 can be rewritten as:

$$P_R = K_A \gamma H \,(1.21) \quad (3)$$

Hossain and Lytton have developed design equations for HDPE manholes that incorporate the effect of placing and compacting the backfill in lifts [1]. The equations give radial and axial (downdrag) loads and deformations. These equations have not yet been confirmed by field measurements. Although they typically yield higher radial pressures than given by Eq. 3, the resulting hoop strain found with Hossain and Lytton's equations is lower than that obtained from the calculations given in this paper when using Eq. 3. This occurs because Hossain and Lytton included the effects of stress relaxation in their equations. Likewise, considerably lower axial stresses are obtained from Hossain and Lytton's equations than when using the equations given herein [4]. (The axial stress is usually the controlling design parameter.)

DOWNDRAG (AXIAL SHEAR STRESS)

The settlement of backfill material surrounding a manhole shaft develops a shear stress between the manhole and the fill, which acts as "downdrag" along the outside of the manhole. The settling process begins with the first lift of fill placed around the manhole and continues until all the fill is placed and consolidated. As fill is placed around a rigid (or otherwise unyielding) manhole, the axial force coupled into the manhole will increase until it equals the static frictional force between the soil and manhole. When this limit is reached, slippage of the fill immediately adjacent to the manhole can occur. Thus, the axial force causing downdrag is limited to the value of the static frictional force.

$$T_A = \mu \left[\frac{P_{R1} + P_{R2}}{2} \right] \tag{4}$$

Where:
- T_A = Average Shear (Frictional) Stress, kPa
- P_{R1} = Radial Earth Pressure at top elevation, kPa
- P_{R2} = Radial Earth Pressure at bottom elevation, kPa
- μ = Coefficient of friction of manhole/soil

The coefficient of friction depends on the manhole material as well as the backfill soil. HDPE manholes are normally installed in granular or cohesive-granular backfill materials. Swan et al. have reported measurements of the angle of friction between HDPE and cohesive granular soil [5], while Martin et al. have reported the same for HDPE and sands [6]. These data and field measurements indicate that a conservative value with regard to downdrag load for the coefficient of friction between HDPE and granular soil is 0.4. The coefficient of friction may be reduced by coating the exterior of the manhole with bentonite or some other lubricant. (It should be noted that the use of external stiffeners or open profiles to stiffen the shaft, greatly increase the downdrag load because any structural component extending out from the manhole shaft will impede the vertical settlement of the soil beside the manhole and thus have the same effect as increasing the coefficient of friction.)

The maximum axial force acting at the bottom elevation of an unyielding manhole can be found by integrating the shear stress (or frictional stress) between the manhole and soil over the height of the fill. The integration is equal to the product of the surface area of the manhole times the average shear stress acting on the surface and is given in Eq. 5. Assuming that the full frictional force is mobilized over the manhole surface, the average shear stress acting on that surface, is given by Eq. 4.

$$P_D = \pi \, T_A \, D_M \, H \tag{5}$$

Where:
- P_D = Downdrag Load, kN
- D_M = Mean Diameter of Manhole, m
- T_A = Average Shear Stress, kPa
- H = Height of Fill, m

This equation can be used for HDPE manholes with the recognition that the HDPE manhole is not an unyielding manhole. Axial deflection of the manhole will lessen the downdrag load. The actual load will depend on the relative stiffness between the manhole and the soil and on the effect of stress relaxation properties on the relative stiffness. The more compliant the HDPE is with the soil properties the less load will be sheared into the manhole. Thus, Eq. 5 will generally over predict the axial load due to frictional shear in the manhole.

GROUNDWATER PRESSURE

The presence of groundwater around a manhole exerts an external hydrostatic pressure on the shaft as well as a buoyant, uplift force on the manhole floor. When calculating the radial pressure acting on the manhole the groundwater pressure should be added to the lateral soil pressure. When soil is submerged, the lateral earth pressure is reduced because the buoyant force reduces the weight of the soil. Eq. 6 can be used to find the lateral pressure in a fully saturated fill surrounding the manhole.

$$P_R' = \gamma_W H + (\gamma_S - \gamma_W) H K_A 1.21 \tag{6}$$

Where: P_R' = Applied Radial Pressure, kPa
K_A = Active Earth Pressure Coefficient
H = Height of Fill, m
γ_W = Unit weight of Water, kN/m³
γ_S = Unit weight of Saturated Soil, kN/m³

For calculating the critical pressure for radial buckling Eq. 6 should be compared with Eq. 3. The highest pressure should be used. For downdrag calculations, the groundwater term in Eq. 6 equals zero because the hydrostatic force is not transmitted in shear. Where groundwater exists, the maximum downdrag force may occur immediately after construction before groundwater can rise or when groundwater falls below the surface grade elevation. In the first of these cases, the pressure is given by Eq. 3. For the second case, where a partial water level exists, the radial pressure equals the sum of Eq. 3 (for the zone above the water level) and Eq. 6 (for the zone below the water level). In this case, H′ as given in Eq. 7 should be substituted for H in Eq. 6.

$$H' = H - Z \tag{7}$$

Where: H′ = Height of fill below groundwater level, m
H = Height of fill, m
Z = Water depth below surface grade, m

The frictional force acting on the manhole with a partial water level equals the sum of the average frictional force due to the radial pressures from the surface elevation to depth Z and the average frictional force due to radial pressures from depth Z to H.

Where manholes are located beneath the groundwater level, the designer should give consideration to restraining the manhole to prevent flotation. The groundwater exerts a force on the manhole equal to the weight of the water it displaces. Restraint is provided by downward resisting forces, which include the weight of the manhole and the downdrag load. However, the full downdrag load given by Eq. 5 may not develop. Therefore, most designers anchor the manhole to a concrete base or ring. When a ring is used, the buoyant weight of the column of soil projecting above the ring can be added to the resisting force. Axial loads in the manhole shaft are minimized by keeping the ring close to the base.

DESIGN PROCEDURE FOR HDPE MANHOLES

The typical manhole consists of the vertical shaft, a floor, a top, and outlets. Each of these components has unique design requirements. The shaft must resist earth and groundwater pressures and carry the weight of live load such as equipment and vehicular loading. The floor has primarily to resist groundwater pressure. The top must transmit live load to the shaft. Consideration must be given to the attachment of outlets so that they do not induce unduly high bending moments or shear stresses into the shaft wall. Outlets attached above the lowest part of the manhole are subject to shear due to fill settlement.

Manhole Shaft Design

Design of the manhole shaft consists primarily of assuming a trial wall section and checking its performance limits for the radial and downdrag loads. Usually, the maximum loads occur near the manhole floor. As loads decrease near the surface, the wall section can be tapered.

RADIAL LOAD

The performance limits under radial load consist of ring compression, ring bending, and ring buckling. Ring compression and ring bending create a combined strain in the manhole wall. The designer limits this strain to an allowable value.

Ring Compression

Radial load acting on the manhole creates a compressive hoop thrust. For a vertical shaft the maximum thrust occurs at the deepest section. (Because of the stiffening influence of the manhole floor, the maximum thrust actually occurs slightly above the floor.) Eq. 8 gives the ring thrust.

$$N_T = P_R\, R_M \tag{8}$$

Where: N_T = Ring Thrust, N/cm
P_R = Applied Radial Pressure, N/cm² (N/cm² = 10 kPa)
R_M = Mean Radius of manhole, cm

For applied radial pressure use Eq. 3, if dry, and Eq. 6, if groundwater is present.

For a given wall section, the ring compressive strain due to the ring thrust is given by Eq. 9:

$$\varepsilon_T = \frac{N_T}{EA_S} \tag{9}$$

Where: ϵ_T = Ring Compressive Strain, cm/cm
N_T = Ring Thrust, N/cm
E = Stress Relaxation Modulus, N/cm²
A_S = Manhole cross-sectional area, cm²/cm
(For solid wall constructions, A_S equals the wall thickness)

The ring strain is proportional to the stress relaxation modulus. For a given HDPE material, the stress relaxation modulus is dependent on the time of loading and temperature. The ring strain calculated by Eq. 9 will be combined with the bending strain from the following section to determine the design adequacy of the proposed wall section.

Ring Bending

Ideally, the radial pressure applied to a manhole acts uniformly around the circumference. This ideal load would not produce a bending moment in the manhole. However, the actual load varies around the circumference as demonstrated by the factor in Eq. 3. This is due to variability in the fill material and its placement. It can be assumed that ring bending deflections will be low, generally on the order of one or two percent of the manhole diameter.

The bending strain can be found either by using an equation that relates the deflection in a cylinder (or pipe) to the strain (such as Molin's Equation) or by the following method which considers the bending moment induced by the eccentricity of the thrust load. The eccentricity factor can be calculated from Eq. 10.

$$e = \mu \ (D_M/2) \tag{10}$$

Where: e = Eccentricity, cm
μ = 0.02 (Ovality correction factor for 2% deflection)
D_M = Mean Diameter of Manhole, cm

The resulting bending moment due to the ring thrust acting over the eccentricity can be found from Eq. 11.

$$M_E = e \ (N_T) \ (0.5) \tag{11}$$

Where: M_E = Bending Load, N-cm/cm
e = Eccentricity, cm
N_T = Ring Thrust, N/cm

The ring bending strain, ϵ_B, for a given section is given in Eq. 12.

$$\varepsilon_B = \frac{M_E}{ES_X} \tag{12}$$

Where: S_X = Section Modulus, $cm^3/cm = I/z$
I = Moment of Inertia of Manhole Wall, cm^4/cm
z = Centroid for Pipe Cross Section, cm
E = Stress Relaxation Modulus of HDPE, N/cm^2

If the stress relaxation modulus for compression is different than the stress relaxation modulus for bending the respective values should be used in Eq. 9 and Eq. 12. (Stress relaxation values may be obtained from the manhole manufacturer or HDPE resin supplier.)

Combined Ring Compression and Ring Bending Strain

For a given trial wall section, the combined ring compression and bending strain should be less than the allowed strain. Strain capacity of HDPE can vary depending on the particular resin, its molecular weight, and its molecular weight distribution.

The maximum combined strain is given by Eq. 13.

$$\varepsilon_C = \varepsilon_T + \varepsilon_B \tag{13}$$

Where: ϵ_C = Combined Strain, cm/cm
ϵ_T = Compressive Thrust Strain, cm/cm
ϵ_B = Ring Bending Strain, cm/cm

In general, the ring strain in a manhole is almost always in compression since strains due to thrust are generally much larger than the bending strain and therefore cancel the tensile component. The long-term, allowable compressive strain is determined by the material's capacity to withstand shear. The allowable strain (sometimes referred to as strain limit) may be determined by accelerated laboratory testing. Because of the variations in HDPE resins and blends, the strain limit should be established for each particular material. Test data for the end-user should be available from the manufacturer.

Practical experience has shown that uneven placement of fill around a manhole and non-uniform settlement of the fill can induce longitudinal bending in the manhole shaft. This bending leads to tensile strains occurring in the axial direction in the manhole. Exceeding a certain level of tensile strain can initiate slow crack growth (SCG), which may lead to macroscopic cracking. Stress-rated PE's have a high strain threshold and resistance to SCG. Blending PE with recycled material or with lower density resins usually reduces this threshold significantly. Since the magnitude of these strains is difficult to predict due to the divergent nature of irregular settlement, stress-rated materials or materials with proven resistance to SCG are normally selected for manholes.

An alternate design approach is to design for stress rather than strain and use an allowable compressive stress value. The limiting stress approach is usually applied to pressure pipe where the pipe is subjected to long-term hoop stress that must be kept below the threshold for developing SCG within the design life. For several years, it was customary to design gravity flow HDPE pipes using an allowable compressive stress approximately equal to the hydrostatic design stress. However, it has recently been shown that the long-term, compressive design stress is higher than the hydrostatic design stress, primarily due to a difference in failure mechanisms [7]. This method can be used by converting the strain in Eq. 13 to a combined stress value.

Ring Buckling

If the radial thrust stress exceeds a critical value the manhole can lose its ability to resist flexural deformation and buckle. Moore and Selig have used continuum theory to develop design equations for buckling [8]. The continuum theory addresses buckling of cylindrical structures surrounded by soil. The presence of groundwater tends to lower the critical buckling value as fluid pressure is not relieved by small deformations that would promote arching in a soil. A solution for hydrostatic pressure effects has not yet been published using the continuum theory. At present the best solution for groundwater effects uses the Winkler model and is given in Eq. 16 [9].

Manhole Section Above Groundwater Level

Moore, Selig, and Haggag give the following equation for the critical ring thrust, N_{CR}, for buried cylinders which can be applied to buried manholes [10]:

$$N_{CR} = 1.2\,\phi\, R_H (EI)^{\frac{1}{3}} (E_S^*)^{\frac{2}{3}} \tag{14}$$

Where:
- N_{CR} = Critical Ring Thrust, (above groundwater), N/cm
- ϕ = Calibration Factor, 0.55 for granular soils
- R_H = Geometry Factor
- E = Stress Relaxation Modulus, N/cm^2
- I = Moment of Inertia of Manhole Wall, cm^4/cm
- E_S^* = $E_S/(1-\nu)$
- E_S = Young's Modulus of the Soil, N/cm^2
- ν = Poisson's Ratio of the Soil

The geometry factor is dependent on the depth of burial and the relative stiffness between the embedment soil and insitu soil. Where the width of the circular zone of fill equals the cylinder radius, the value of R_H approaches unity as the relative stiffness between the manhole and the soil approaches 0.005. Relative stiffness is defined as:

$$Relative\ Stiffness = \frac{EI}{G_S r^3} \tag{15}$$

Where:
- G_S = Shear Modulus of Soil Embedment, N/cm^2
- r = Radius of Cylinder, cm

For almost all HDPE manholes installed in a granular or compacted, cohesive-granular embedment, the relative stiffness will be less than 0.005 and R_H equals 1.0. Moore also shows that for deep burial in uniform fills R_H equals 1.0.

For design purposes, the ring thrust as given by Eq. 8 should not exceed one-half the critical ring thrust, N_{CR}.

Manhole Section Below Groundwater Level

The critical thrust, N_{CRW}, for buckling beneath the groundwater level can be determined using Eq. 16. [11]

$$N_{CRW} = 2.825 \sqrt{\frac{RB' E' EI}{D_M}} \tag{16}$$

Where:
- N_{CRW} = Critical Ring Thrust (below groundwater), N/cm
- D_M = Mean Diameter, cm
- R = 1-.33H'/H, Buoyancy Reduction Factor
- H' = Height of Groundwater above Invert
- H = Height of Fill, m
- E' = Modulus of Soil Reaction, N/cm²
- E = Stress Relaxation Modulus, N/cm²
- I = Moment of Inertia of Manhole Wall, cm^4/cm

and:

$$B' = \frac{1}{1+4e^{(-0.213H)}} \tag{17}$$

For design purposes, the ring thrust as given by Eq. 8 should not exceed one-half the critical ring thrust, N_{CRW}.

When radial stiffeners are provided in the manhole wall, the average moment of inertia of the wall can be used in the above equations. But, a check should be made to insure that the spacing between stiffeners does not permit local buckling.

AXIAL LOAD PERFORMANCE LIMITS

In the above section on Earth Loading, the axial load due to downdrag was given. In addition to the downdrag, other axial loads include the weight of the manhole and its appurtenances and the weight of any live loads such as equipment or vehicles. These loads create an axial, compressive strain in the manhole wall. The strain is limited by the compressive strain capacity of the material and by axial buckling. Both limits are calculated and the smallest allowable strain controls design.

<u>Axial Strain</u>

The axial, compressive strain, ϵ_A, resulting from the downdrag force is given by Eq. 18.

$$\varepsilon_A = \frac{P_D+P_l+P_W}{E\pi D_M t_n} \tag{18}$$

Where:
- ϵ_A = Axial Compressive strain, cm/cm
- P_D = Downdrag Force from Eq. 5, N
- P_l = Live load, N
- P_W = Dead load, N
- E = Stress Relaxation Modulus, N/cm²
- D_M = Mean Diameter of Manhole, cm
- t_n = Net Wall Thickness, cm

Some HDPE manholes are fabricated from profile cross-sections. The net wall thickness is the thickness of manhole wall profile minus the width of any hollow geometric core in the profile structure.

For design, the axial strain must be less than the allowable strain for the manhole material.

Axial Buckling

As the axial strain is increased in a cylindrical tube, supported by soil, the tube is subject to local buckling rather than column buckling. In the lowest (local) buckling modes, the tube will deflect outward slightly and dimple inward. For a buried manhole, the resistance to buckling in this manner is increased by the surrounding soil, which acts to restrain outward deflection. Buckling equations for a cylindrical tube with no soil support are given in the literature. These equations can be used for manhole design and are conservative in cases where the surrounding soil is a stable, well-compacted granular material.

One such equation is given by Timoshenko and Gere [12]. It can be restated in terms of the critical strain as given below.

$$\varepsilon_{CR} = \frac{2S_E}{D_M\sqrt{3(1-\mu^2)}} \tag{19}$$

Where:
- ϵ_{CR} = Critical Axial Strain, cm/cm
- D_M = Mean Diameter of Manhole, cm
- μ = Poisson's Ratio of HDPE
- S_E = Equivalent solid wall thickness, cm

$$S_E = \sqrt[3]{12I}$$

Where:
- I = Moment of inertia of wall cross-section, cm^4/cm

For the design of buried manholes this equation can be applied without a safety factor as the soil support will provide sufficient safety factor and the axial loads on a viscoelastic manhole are believed to be considerably lower than predicted by the method given herein [4]. (Where soil support is minimal, such as in saturated loose or saturated fine grain material, an appropriate safety factor should be applied to Eq. 19.)

Chau and Chau et al. have reported on an analytical method for calculating buckling resistance of buried horizontal cylinders subject to axial loads [9,13]. This method may be used for manhole design.

Interaction of Axial and Radial Buckling

The work of Chau shows that the axial loading reduces the critical radial buckling stress only slightly [13]. This observation is supported by elastic stability methods. (See Timoshenko and Gere [12]).

MANHOLE BASE/FLOOR DESIGN CONSIDERATIONS

The critical load acting on the base (or floor) of a manhole is the groundwater pressure. Many manholes have flat floors that are essentially circular plates. Circular plates generally have low resistance to deflection and stress. Therefore, if there is any potential for groundwater, the designer should check the manhole floor. Typically, the deflection controls the floor design. Most manhole floors are limited to a deflection not greater than two percent for 150 cm (60") and smaller diameter and not greater than one percent for larger diameters. Larger deflections may be tolerable but pumps or other equipment located on the floor can become unstable.

Solutions for stress and deflection of circular plates are given in the literature. Sealy and Smith give the following [14]:

$$\sigma = \frac{3}{4} p \frac{r^2}{t^2} \tag{21}$$

Where: σ = Maximum Stress, N/cm²
p = Groundwater Pressure, N/cm²
r = Radius of Plate, cm
t = Plate Thickness, cm

$$\delta = \frac{3}{16}(1-\mu^2)\frac{pr^4}{Et^3} \tag{22}$$

Where: δ = Maximum Deflection, cm
μ = Poisson's Ratio
p = Groundwater Pressure, N/cm²
r = Radius of Plate, cm
t = Plate Thickness, cm
E = Stress Relaxation Modulus, N/cm²

Stiffening gussets can be added to the manhole base to reduce stress and deflection. An analysis should be made to prove that these stiffeners are adequate and that the shear stress in the weld between the stiffeners and the base is acceptable. Manhole bases that are not flat plates may be considered on the basis of more sophisticated analysis or physical testing.

MANHOLE TOPS

For applications subject to H-20 traffic loading, a concrete cap is normally placed over the manhole or the HDPE manhole top is cast in concrete. Although HDPE tops can be designed to withstand the weight of H-20 loads, traffic loads can cause significant deflection of the top and the shaft. The deflection may not damage the HDPE, but it may lead to severe cracking of pavement. Before accepting an HDPE top for street installation, the designer is advised to seek test data from the manufacturer showing the limits of deflection under H-20 loadings.

SUMMARY AND CONCLUSIONS

A design methodology has been described for HDPE manholes. The method gives conservative designs for a wide range of applications. A more complete analysis for HDPE manholes will be available upon completion of field verification of Hossain and Lytton's equations.

REFERENCES

[1] Hossain, M.K. and Lytton, R.L., "Analysis of Large Diameter High-Density Polyethylene Plastic Pipes as Vertical Shafts in Landfills," Journal of Testing and Evaluation, ASTM, Vol. 19, No. 6, Nov. 1991, pp. 475-484.

[2] Gartung, E., Prühs, H., and Hoch, A., "Design of Vertical Shafts in Landfills", Second International Landfill Symposium, Sardinia, 1989.

[3] Report to Bauku, Wiehl, Germany from Grundbauingeniure Steinfeld und Partner, Erobaulaboratium Hamburg, 1991.

[4] Hossain, M.K., "Finite Element Analysis and Design of Large Diameter Flexible Vertical Pipes Subjected to Incremental Compacted Backfill Loads and Creep Effects," Master Thesis, Texas A&M University, 1990.

[5] Swan Jr., R.H., Bonaparte, R., Bachus, R.C., Rivette, C.A., and Spikula, D.R., "Effect of Soil Compaction Conditions on Geomembrane-Soil Interface Strength, Geotextiles and Geomembranes, 10, 1991, pp. 523-529.

[6] Martin, J.P., Koerner, R.M. and Whitty, J.E., "Experimental Friction Evaluation of Slippage Between Geomembranes, Geotextiles and Soils", International Conference on Geomembranes, Denver.

[7] "The Influence of New Material Strength Data on Product Calculation", BAUKU Symposium, Wiehl, Germany, 1992.

[8] Moore, I.D. and Selig, E.T., "Use of Continuum Buckling Theory for Evaluation of Buried Plastic Pipe Stability," Buried Plastic Pipe Technology, ASTM STP 1093, ASTM, Philadelphia, 1990.

[9] Chau, M.T., Chua, K.M., and Lytton, R.L., "Stability Analysis of Flexible Pipes: A Simplified Biaxial Buckling Equation," 68th Annual Meeting, Transportation Research Board, Washington, D.C., 1989.

[10] Moore, I.D., Selig, E.T., and Haggag, A., "Elastic Buckling Strength of Buried Flexible Culverts," TRB Session 143, Transportation Research Board, Washington, 1988.

[11] Cagle, L.L. and Glassock, B., "Recommendations for Elastic Buckling Design Requirements for Buried Flexible Pipe", Proceedings: "Better Water for The Americas", Part 1, AWWA, 1982.

[12] Timoshenko, S.P. and Gere, J.M. Theory of Elastic Stability, McGraw-Hill Company, 1961, p. 465.

[13] Chau, M.T., "Stability Analysis of Buried Flexible Pipes: A Biaxial Buckling Equation," Master Thesis, Texas A&M University, 1990.

[14] Sealy, F.B. and Smith, J.O., Advanced Mechanics of Materials, John Wiley & Sons, Inc., New York, 1952.

Dennis E. Bauer, P.E.[1]

ORIENTED PVC PIPE (PVCO): EXPERIENCE AND RESEARCH

REFERENCE: Bauer, D. E., "Oriented PVC Pipe (PVCO): Experience and Research," Buried Plastic Pipe Technology: 2nd Volume, ASTM STP 1222, Dave Eckstein, Ed., American Society for Testing and Materials, Philadelphia, 1994.

ABSTRACT: As Polyvinyl Chloride (PVC) sanitary and storm drain pipes have evolved with the advent of profile walls, so too will PVC pressure pipe with the advent of orientation. This paper discusses the basic concepts of reorienting PVC pipe, the basic production technologies involved, and the research and experience which span the past two decades around the world.

The principal benefit of reorientation is increased tensile strength resulting in an increase in the Hydrostatic Design Basis (HDB). However, the improvements in fatigue resistance, impact resistance and crack growth resistance are offshoots of the orientation process which perhaps offer the most significant advantages for end-use applications. The practical affects of these material property improvements are longer life under dynamic pressure conditions and better handling ability than conventionally extruded PVC pipe.

KEY WORDS: plastic pipe, oriented Polyvinyl Chloride (PVC), PVC pressure pipe, PVCO

INTRODUCTION

Oriented PVC (PVCO) pipe is PVC pressure pipe which attains a relatively high strength by reorienting the molecular structure. Whereas conventionally extruded PVC pressure pipe has a maximum hydrostatic design basis (HDB) of 4000 psi, PVCO has a maximum HDB of 7100 psi. In addition to superior tensile strength, the reorientation enhances impact and fatigue resistance.

---

[1]Eastern Regional Sales Manager, Extrusion Technologies, Inc., a Uponor Group Company, 1600 Stout Street, Suite 1710, Denver, CO 80202-3134.

Reorientation is accomplished by expanding (blowing up) conventionally extruded PVC pipe. A conventionally extruded PVC pipe of approximately half the diameter and twice the wall thickness of finished PVCO is placed inside a mold. The temperature is raised to the appropriate level and internal pressure is applied. The pipe is thus blown up to the inside dimensions of the mold (for example, 2" diameter is expanded to 4" diameter). An integral bell is formed within the mold.

MOLECULAR ORIENTATION

PVC molecules are long flexible chains which may be oriented during the manufacturing process. Processing techniques [1], [4] which purposely generate molecular orientation have been applied for several decades to PVC due to the enhanced material properties which may result. Grocery bags, which have extremely high unit tensile strength, and plastic bottles are commonly accepted examples of enhanced product properties through orientation.

In the case of PVCO [2], the molecular structure of conventionally extruded PVC pipe, which is a random arrangement of long chain molecules (Fig. 1), is reoriented into a much less random structure. The molecules become much more closely aligned with each other in the direction of orientation (Figs. 2 and 3).

In order to begin the reorientation process, the conventionally extruded pipe starting stock must be heated to a temperature above PVC's glass transition temperature, approximately 80°C (175°F), when it passes from a rigid "glassy" state into a rubbery state. Deformation (blowing up) in the rubbery state causes molecular orientation between the cross links. The optimum temperatures for orientation lie between the glass transition temperature and approximately 120°C (250°F).

Once the starting stock is placed in the mold, hot water is circulated both in and outside the pipe to bring the pipe temperature to the desired level. At that point the pressure on the inside of the pipe is raised to expand or inflate the pipe. Simultaneously the heated water on the outside of the pipe is evacuated allowing the pipe to expand to the limits of the mold. The heated water is then exchanged for chilled water which "freezes-in" the reoriented structure and its accompanying property enhancements. The PVC pipe is then removed from the mold.

RESEARCH

Research into the enhanced properties and production techniques for PVCO pipe span the past two decades [1]. Currently research is being conducted primarily by Uponor, headquartered in Finland, and Vinidex Tubemakers Pty Ltd of Australia (Vinidex).

HDB: As with conventionally extruded PVC pipe, an HDB is determined by selecting a long term hydrostatic stress (LTHS) from the stress regression line developed for a specific compound. Figure 4 represents typical results of a PVCO stress regression compared to that of conventionally extruded PVC. PVCO's line parallels that of PVC and reveals that a hoop stress of approximately 1.75 to 2.0 times greater is required to burst PVCO. Therefore, when identical safety factors are applied, PVCO pipe requires less wall thickness than PVC pipe when both are pressure rated the same.

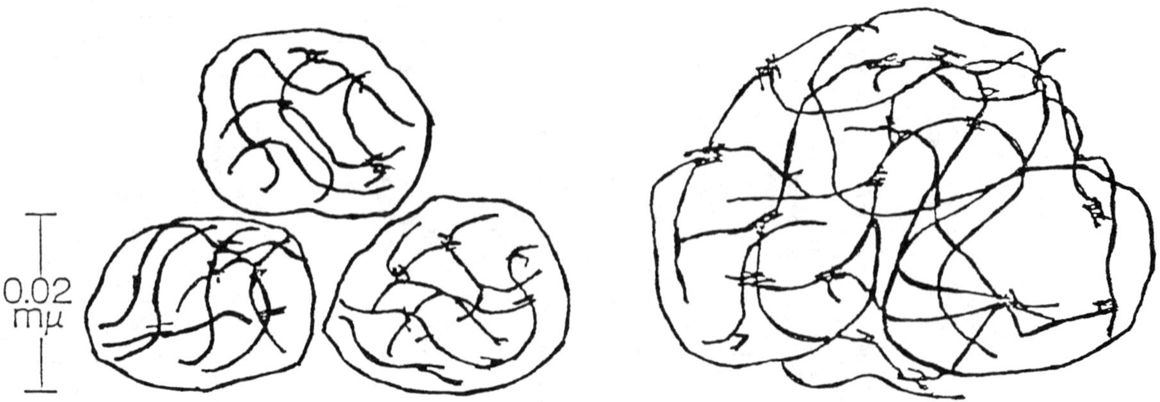

Figure 1. Conventionally Extruded PVC Molecular Structure

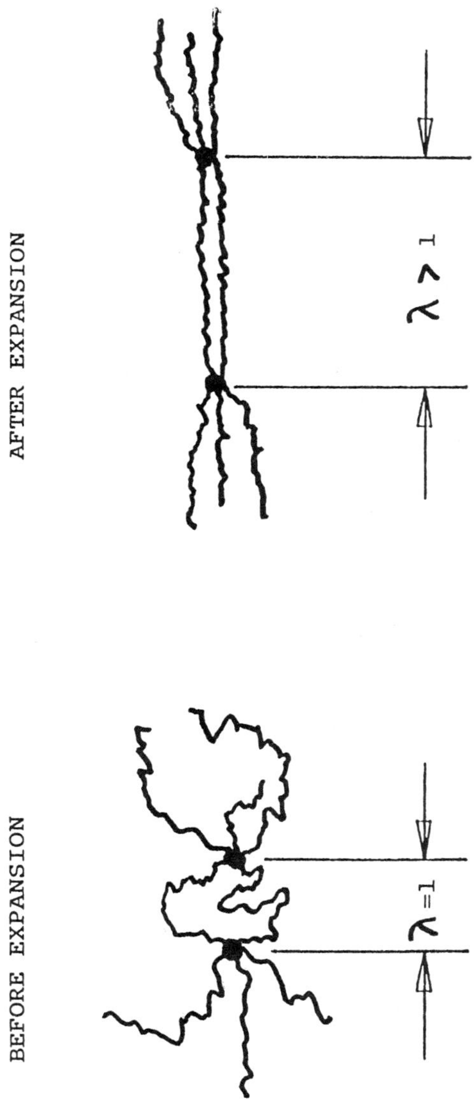

Figure 2. Molecular Structure Before and After Reorientation

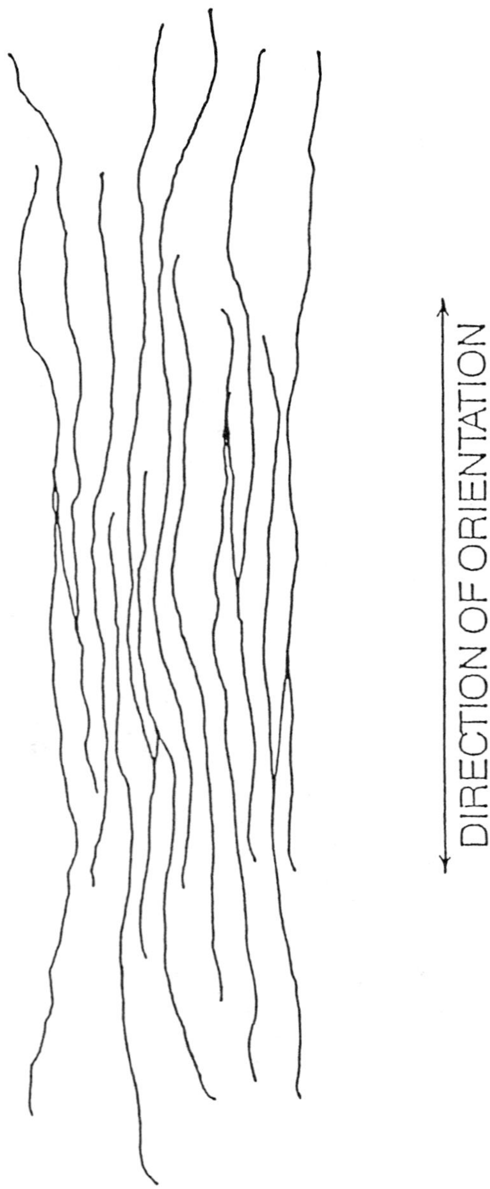

Figure 3. Oriented PVC (PVCO) Molecular Structure

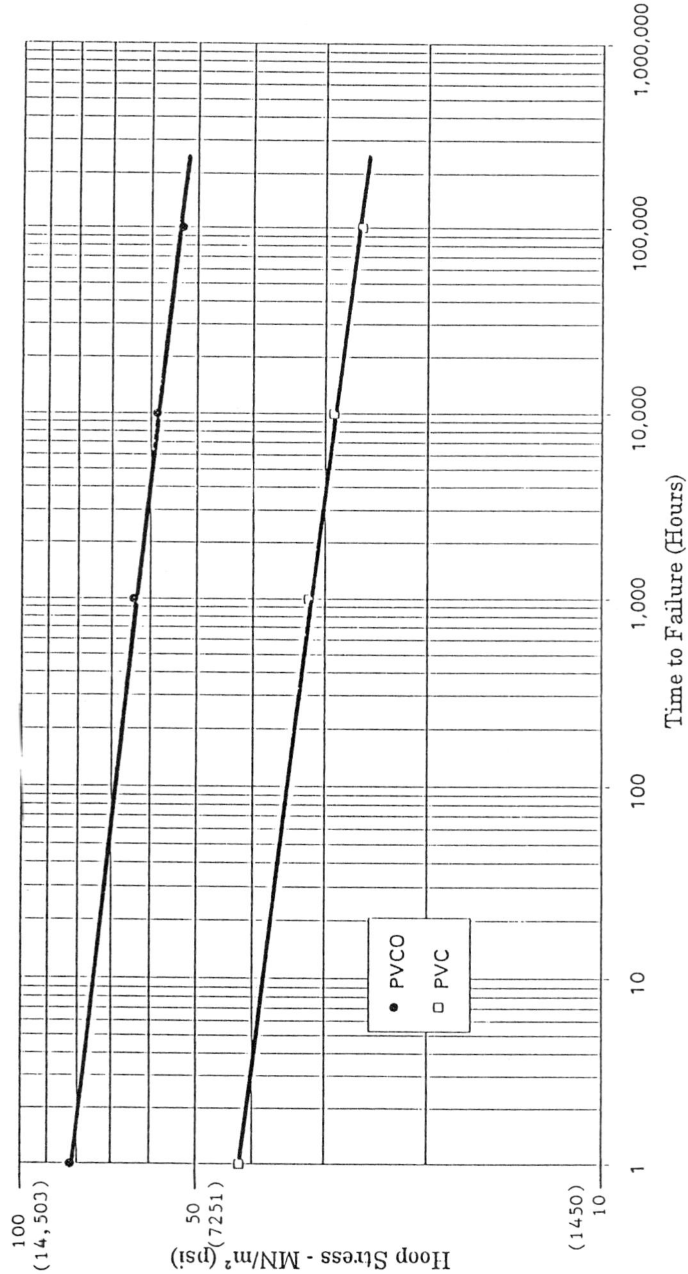

Figure 4. Typical Long Term Hydrostatic Pressure Comparison of PVCO and PVC

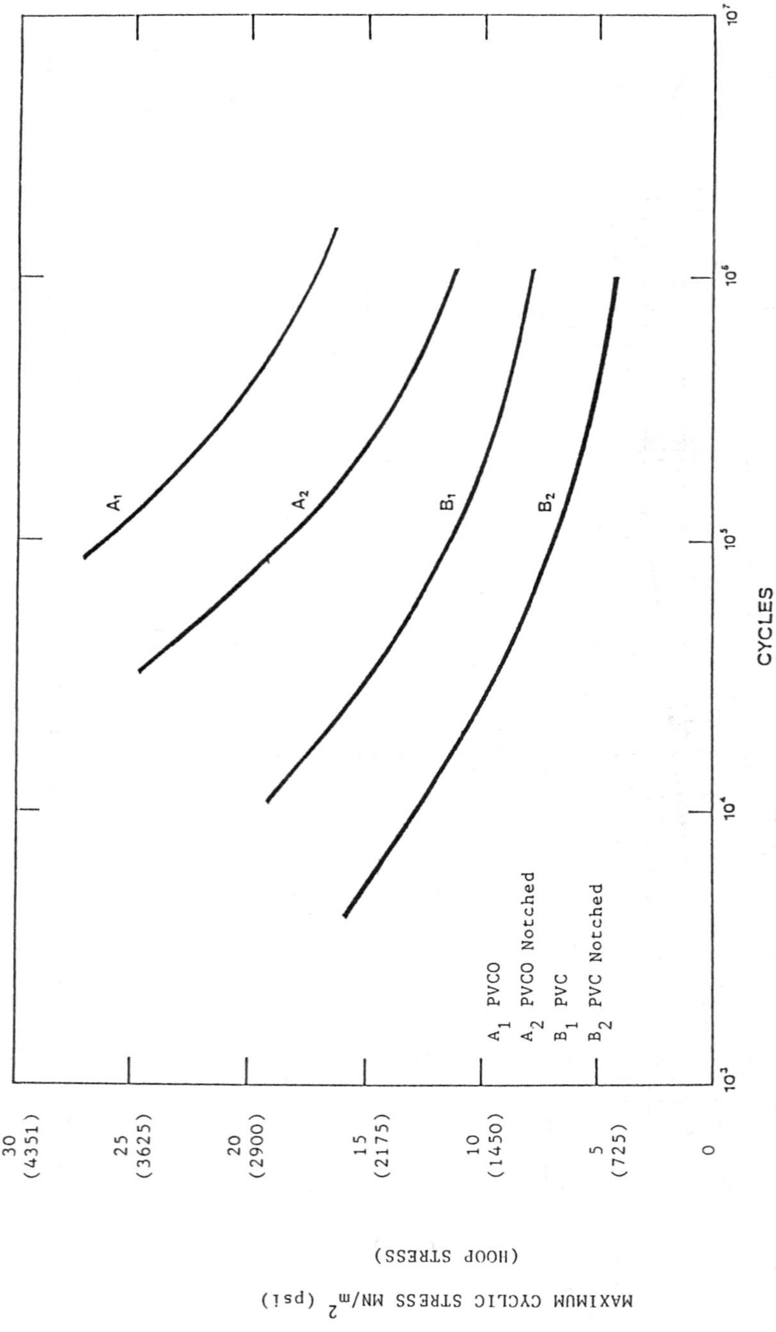

Figure 5. Notched and Plain Mechanical Fatigue Data

UPONOR

IMPACT ENERGIES
H.S.P.V.C. V. '1ST GENERATION' PVC-U

PIPE	ENERGY TO FAILURE (O°C) (kgfm)
HSPVC	11.7
PVC-U PIPE TO BS 3505 (SAME WALL THICKNESS AS HSPVC)	1.5
PVC-U PIPE TO BS3505 (SAME PRESSURE RATING AS HSPVC)	2.3
AND AFTER COMPREHENSIVE TESTING OF 6IN. DIAMETER HSPVC PIPE THE FOLLOWING DATA WAS GENERATED:	
HSPVC	60.0
PVC-U (SAME WALL THICKNESS)	3.4

Impact Strength: Falling weight impact tests both at room temperature 20°C(68°F) and at 0°C (32°F) have demonstrated substantial improvements in failure level of PVCO pipe compared with PVC manufactured to ASTM D2241, "Standard Specification for Poly(Vinyl Chloride) (PVC) Pressure-Rated Pipe (SDR Series)," or British Standard BS3505, "British Standard Specification for Unplasticized Polyvinyl Chloride (PVC-U) Pressure Pipes for Cold Potable Water." Testing at 0°C demonstrates that PVC has from 3 to 5 times greater impact capability than PVC, of the same pressure rating, made to D2241 and BS3505 respectively.

Fatigue Resistance: Figure 5 reveals the test data comparing PVCO and PVC, for cycles to failure versus cyclic stress. This graph demonstrates that PVCO can sustain substantially more cycles for a given maximum cyclic stress than PVC. From these data, PVCO demonstrates roughly 50 times greater resistance. Or, given a specified number of cycles, PVCO can sustain a much higher maximum cyclic stress than PVC: approximately two times greater.

This mechanical stress work was conducted by Professor P.P. Benham at Queens University, Belfast.

Typical applications where fatigue becomes a critical design variable are force mains or golf course irrigation systems where pumps come on and off, or valves open and close, at frequent intervals.

Stress Cracking (Crazing): Another very positive property enhancement resulting from orientation is dramatically increased resistance to crazing [2,3]. Orientation of samples prior to tensile tests, results in steeper elastic deformation curves than normally extruded PVC, a higher yield stress, a lower reduction in stress during yielding and less necking of the samples. At a given orientation level (λ=1.75) the yield point disappears, there is no drop in stress and no necking of the sample will take place. This means the possibility of local yielding, which is necessary for the forming of fibrils in the crazes, has disappeared and craze initiation is no longer possible. Therefore, PVC is much more likely to exhibit a ductile failure mode rather than a premature brittle failure. This provides greater safety and certainty about the lifetime of PVCO pipes.

EXPERIENCE

Yorkshire Imperial Plastics (YIP) Ltd. first developed PVCO pipes under the trade name *"Superpolyorc,"* following original work carried out by Imperial Chemical Industries PLC. Extensive development was undertaken by YIP, beginning in 1972, and the product proven in diameters ranging from 100 mm (4") to 450 mm (18"). The first installations were made in the U.K. in 1974. Full scale production, beginning with 150 mm (6") diameter, began in 1982. In 1987, YIP was acquired by Uponor and full scale production of 200 mm (8") began. Over the past fifteen years, several hundred thousand feet of 150 mm and 200 mm (6" and 8") PVCO have been installed throughout Europe.

Vinidex began production and delivery of PVCO in 1984 in Australia. Since then, over two hundred kilometers (over a half million feet) have been installed throughout Australia.

Extrusion Technologies, Inc. (ETI) began installing test projects across the U.S. in the spring of 1991. Full scale production of 150 mm and 200 mm (6" and 8") PVCO began in late 1992. By the time this paper was written, several thousand feet had been installed within the U.S. Plans are currently underway to increase the diameter range.

The experiences with PVCO have been very favorable. Because it is expanded under a uniform internal pressure, both the internal and external finish of the pipe are extremely smooth. While primarily an aesthetic quality, it is one which is highly thought of by the end user. Because the outside diameter is the same as the existing products in the local, the fitting availability and requirements are identical to currently specified PVC products. Therefore, there are no special items required for PVCO. Because the O.D. is the same as existing PVC, yet the wall thickness is reduced, PVCO has a larger I.D. which allows for greater flow capability for a given diameter. Lighter weight and increased impact resistance have meant greater handleability. PVCO cannot be solvent cemented. However, the gasketed bell and spigot joint are preferred for most underground applications due to ease of assembly.

CONCLUSIONS

PVCO pipe offers several product property enhancements which are beneficial for its intended applications as a pressure pipe. Increased tensile strength, impact resistance, fatigue resistance and resistance to crazing are all significant improvements to a product which has performed extremely well for over fifty years - PVC pipe.

At a time when the global economy dictates efficiency, quality and performance, innovations such as PVCO deserves our evaluation and acceptance.

REFERENCES

[1] Richard C. Stephenson, "Some Recent Developments in the Processing of Poly(Vinyl Chloride)," Journal of Vinyl Technology, Vol. 5, No. 1, 1983.

[2] T. G. Meijering, "The Production and Application of Bioriented PVC Pipe," Plastics Pipe VII, Book of Proceedings, Europe, 1990.

[3] Dr. Ir. L.C.E. Struik, "Crazing & Fracture," TNO report, Nov. 1981.

[4] T. G. Meijering and P. Benjamin, "Moleculaire Biorientatie van PVC/CPE Gazbuizen," Gas 9, 1986.

Rehabilitation

Douglas G. Kleweno[1]

PHYSICAL PROPERTIES AND CHEMICAL RESISTANCE OF SELECTED RESINS FOR CURED-IN-PLACE PIPE REHABILITATION

REFERENCE: Kleweno, D.G., "Physical Properties and Chemical Resistance of Selected Resins for Cured-in-Place Pipe Rehabilitation," Buried Plastic Pipe Technology: 2nd Volume, ASTM STP 1222, Dave Eckstein, Ed., American Society for Testing and Materials, Philadelphia, 1994.

ABSTRACT: This report is a summary of the basic chemistry, physical properties, and chemical resistance of six commercially available resins (i.e. vinyl ester, polyester, and epoxy) used for making cured-in-place pipe (CIPP). Flexural and tensile properties are included and chemical resistance from a one year study for each resin system as a CIPP resin/felt composite in acids, bases, and oxidizing agents currently found or introduced into municipal sanitary sewer systems [1,2,3,4].

Conclusions drawn from this report indicate the three different types of resin had measurably different flexural and tensile properties that ranged the from stiff, brittle behavior to very flexible. Among the test group the two polyesters were relatively stiff, the two epoxy vinyl esters had a balance of high stiffness and strength, one epoxy system was extremely flexible, while the other was similar to the epoxy vinyl esters. The one year chemical resistance also clearly distinguished the performance of the three types of resins. Overall, the two polyester resins performed at an intermediate level, the two epoxy vinyl ester resins demonstrated superior broad range chemical resistance, and the epoxy performance ranged from poor to excellent depending on the type of curing agent used.

KEY WORDS: physical properties, chemical resistance, thermoset resins, cured-in-place pipe, pipe

Accelerated aging caused by hydrogen sulfide-related corrosion has generally caused premature failure of our nation's sanitation infrastructure. Awareness of the existence of corrosion and concern about its effect on the sewer system has been an issue since concrete first started displacing clay and brick as the primary materials of sewer construction. Even though it was known that some corrosion would take place, precautions taken in the sewer design and pipe thickness were intended to produce the 100+ year life expectancy of the sewer system [1]. However, within the last 15 years, hydrogen sulfide-related

[1]Sr. Development Engineer, Chemical Resistance Application Development Labs, The Dow Chemical Company, Texas Division, 2301 N. Brazosport Blvd., B-2009, Freeport, Texas 77541-3257

corrosion has accelerated at an alarming rate throughout the U.S. and has been documented by the Environmental Protection Agency (EPA) in a number of studies [1,2,3,4]. The primary cause of the accelerated corrosion has been attributed to the proliferation of several strains of Desulfovibrio bacteria in response to the reduction of cyanide and other heavy metal pollutants regulated by the EPA [1,2]. An anaerobic bacteria living in the slime layer on the lower hemisphere of the pipe reduces sulfur-containing compounds to hydrogen sulfide (H_2S). An aerobic strain living in the slime on the crown of the pipe oxidizes hydrogen sulfide to sulfuric acid (H_2SO_4). The aerobic bacteria have been observed to produce sulfuric acid up to 5% by weight (i.e., pH $\simeq$ 0.28) and remains viable in concentrations as high as 7% (i.e., pH $<$ 0.15) [2,5].

Sewage overflow restrictions, overflow monitoring, and stiff fines for noncompliance imposed by the EPA and state water agencies have motivated municipal sanitation departments to develop aggressive programs to maintain and/or rehabilitate their systems [2]. These programs have fostered the growth and acceptance of number of trenchless pipe rehabilitation techniques, as well as creative maintenance solutions [3,4,6].

The current emphasis on rebuilding infrastructure has created the environment for the development and acceptance of a number of different methods of performing CIPP. Although the construction and installation of each technique varies slightly, they all incorporate the use of non-woven and/or woven fabric formed into a tube that is impregnated with a thermosetting resin and subsequently cured in place to form a new pipe inside the old conduit. Once the job is completed the client has a new pipe composed of approximately 65-85% thermosetting resin and 15-35% fibers or fabric. Assuming consistent installation, the long term performance of that product will be strongly controlled by the type of resin used, no matter what fibrous material was used to construct the tube itself.

In order to extend the life of existing intact concrete sewer systems, many sanitation districts have developed chemical treatment protocols and inventive application techniques to control hydrogen sulfide corrosion [2,4,6]. Depending on the program objectives, regular addition of one or more chemicals can reduce existing hydrogen sulfide, neutralize the acids, temporarily shock the bacteria, or accomplish all three. Chemicals commonly used for this purpose includes strong oxidizing agents (i.e., hydrogen peroxide, sodium hypochlorite (active ingredient in bleach), chlorine, potassium permanganate), weak oxidizing agents (i.e., oxygen and air injection), acid neutralizing bases (i.e., sodium hydroxide), and iron salts [2,4].

In general, this nation's sanitation system has changed dramatically within the last two decades and will continue to evolve. Studies demonstrate that decreased flows increase the corrosive environment in sewer systems [7]. It is suggested that municipal efforts to reduce inflow and infiltration (I/I) through rehabilitation will also increase hydrogen sulfide-related corrosion and concentrate all other chemical agents present. These and other unpredictable changes may necessitate chemical treatment within sanitation systems that presently are not chemically treated. However, most chemicals used for hydrogen sulfide reduction can be much more chemically aggressive to general purpose thermoset polymers than hydrogen sulfide or sulfuric acid. To date, only municipalities using the "Green Book" require chemical resistance qualification for CIPP that will be exposed to chemicals introduced into the sanitation system [8,9].

Cured-in-place pipe has benefitted tremendously by the development of ASTM standardization and contractor specific documentation detailing the method of construction and installation. However, a review of Standard Practice for Rehabilitation of Existing Pipelines and Conduits by the Inversion and Curing of a Resin-Impregnated Tube (F 1216) and most municipal material of construction specifications for CIPP offer the municipal wastewater engineer very little guidance for the classification, relative performance, or selection of the thermosetting resins used for CIPP. As a result, most municipal wastewater engineers are left with no alternative but to rely exclusively on the contractor to select the thermosetting resin. The civil engineering community specifying CIPP needs this information and must become more knowledgeable. Proper screening and selection of the resin systems used for all aspects of sewer rehabilitation will maximize use of project funds for the present conditions of the sewer and may also provide insurance against anticipated and unanticipated environmental changes that will be occurring in the next 100 years [9]. The data in this paper is intended as one general source of information about thermosetting resins used for cured-in-place pipe.

EXPERIMENTAL METHODS

Materials

The resins used in this study are all commercially available and have been or are currently being used for CIPP rehabilitation.

1. Isophthalic polyester resin-1 (PER-1)
2. Isophthalic polyester resin-2 (PER-2)
3. Bisphenol A epoxy resin-1/Flexibilized amine curing agent (ER-1)
4. Bisphenol A epoxy resin-2/Multiamine/Imidazole curing agent (ER-2)
5. Bisphenol A epoxy vinyl ester-1 (VER-1)
6. Bisphenol A epoxy vinyl ester-2 (VER-2)

The PER-1, PER-2, VER-1, and VER-2 resins were catalyzed with a cobalt promoter and the combination of two peroxide initiators. The same bisphenol A (bis A) epoxy resin was cured with two different amine based curing agents according to proper reactive equivalent stoichiometric ratios. Cured-in-place pipe composite samples made with these six resins were composed of resin impregnated polyester felt that was vacuum impregnated with catalyzed resin and cured 4.5 hours at 100°C (212°F) between spaced aluminum plates. The polyester felt consists of short polyester fibers needled into a dense felt. Polyester felt was chosen because it is currently the most common material of construction for CIPP rehabilitation.

Test Methods

Three point bend flexural testing was carried out according to the ASTM Test Method for Flexural Properties of Unreinforced and Reinforced Plastics and Electrical Insulating Materials (D 790) test method I. A standard L/d ratio of 16:1 was used and specimen dimensions, support span, and rate of loading was determined from Table 1 in D 790. Specimen dimensions were held constant at 12.7 mm x 3.2 mm x 76.2 mm and tested with a 50.8 mm (2 in.) support span at a rate of 1.3 mm/min (0.05 in/min). Tangent modulus of elasticity, maximum stress and strain were determined as described by test method I conditions within D 790. Tensile properties were evaluated according to ASTM Method for Tensile

Properties of Plastics (D 638). Specimens were approximately 3.175 mm (0.125 in.) thick and were dog boned to conform with Type I dimensions given in Fig. 1 of D 638. A 50.8 mm (2 in.) extensometer was attached to the narrow section of each specimen for accurate measurement of percent elongation and slope. Specimens were loaded at a constant rate of 5.08 mm/min (0.2 in./min.). All specimens broke in a brittle manner without any yield and this is shown in Fig. A1.2 of D 638. Modulus of elasticity, strength and elongation at break were determined as described in D 638. All flexural and tensile specimens were equilibrated at 70°F and 50% relative humidity for 40 hours prior to testing. Testing was done with an Instron load frame (model 4505) equipped with auto-ranging 1 kN (224 lbf) and 50 kN (11.2 kip) load cells for flexural and tensile testing, respectively.

The chemical resistance of CIPP composites were evaluated in extremes of acidity and basicity, as well as to oxidizing agents. The acids tested are 25% sulfuric, 5% nitric, and 20% hydrochloric (having low pH), the bases are 5% ammonium and 5% sodium hydroxide (having high pH), and the oxidizing agents are 0.5% and 5% sodium hypochlorite (i.e., bleach), 5% hydrogen peroxide, and 5% potassium permanganate. The chemical baths were maintained at 49°C (120°F) since this is within the 140°F input temperature limit frequently enforced by most sanitation departments. Chemical resistance testing was accomplished with 3.175 mm (0.125 in.) thick resin/polyester felt composites having exposed edges and a nominal surface area of 5806 mm^2 (9 $in.^2$). Chemical exposure and coupon evaluation was carried out in a manner consistent with the ASTM Standard Practice for Determining Chemical Resistance of Thermosetting Resins Used in Glass-Fiber-Reinforced Structures Intended for Liquid Service (C 581). Once removed from the chemical baths, the coupons were rinsed with tap water, and patted dry with paper towels before measuring weight and thickness. The coupons were then cut and evaluated for flexural properties as previously described. Chemical resistance coupons were not equilibrated, but were tested under conditions of 70°F and 50% relative humidity within 24 hours upon removal from the chemical bath. The majority of specimens were evaluated at intervals of 1, 3, 6, and 12 months. Some of the ER-1, PER-1, and PER-2 specimens prematurely failed in some of the chemicals before the 12 month evaluation. Only 6 months of data is currently available for the ER-2 resin system because it was started at a later time and the study has not been completed.

DISCUSSION AND RESULTS

Resin Chemistry and Processing

Introduction--The three types of thermosetting resins most commonly used for cured-in-place pipe are polyester, vinyl ester, and epoxy resins. Polyesters are the least expensive and are by far the most commonly used thermoset with vinyl esters and epoxy resins used to a lesser degree. Just as "concrete" is a generalized name for that material of construction, polyester, vinyl ester, and epoxy are generic names for a wide variety of resins that fall into each category. In addition, just as aggregate size, shape, and volume might determine the price and performance of concrete, the molecular components of thermosetting resins also determine price and performance.

Polyester--Unsaturated thermosetting polyester resins are the product of a polycondensation reaction between specific glycols and saturated and unsaturated dicarboxylic acids. Typical glycols, saturated and unsaturated dicarboxylic acids have been listed in Figure

1. It is the saturated acid by which the polyester resin derives its name and an element of its inherent performance. The unsaturated dicarboxylic acid contributes the reactive polymerization site. The glycol links the two types of dicarboxylic acids together. A common (although not the only) class of polyesters used for CIPP is that produced with isophthalic acid, and is therefore called an unsaturated isophthalic (Iso) polyester resin. The generic structure of an Iso polyester is given in Figure 2 where a number of the predominant features of the polymer have been labelled. The term "unsaturated" refers to the (-C=C-), which are the polymeric reaction sites that facilitate cross linking during the curing process. The multiple ester groups (-COO-) are the weak link in the polymer, providing sites for hydrolysis in the presence of strong acids and bases.

Epoxy--Epoxy resins are the reaction product of various types of phenols and epichlorohydrin, which provides the polymeric reaction site. The type of phenol used is typically how the name of the epoxy is derived. The most common epoxy is that made with bisphenol A (bis A) and is often referred to as a bis A epoxy resin (Fig. 4). A listing of some of the most common phenols are given in the flow chart in Figure 3. The bis A epoxy backbone (Fig. 4) provides toughness and the polymeric reaction sites are only located at each end of the molecule. Hydroxyl (-OH) groups in the middle and hydroxyl groups produced at the reaction site can provide improved adhesion (Fig. 4). However, the polar hydroxyl groups, choice of curing agent, and/or presence of reactive diluents (see Processing) are also sources for chemical degradation.

Vinyl Ester--Epoxy vinyl ester resins are the reaction product of an epoxy resin with methacrylic acid. The methacrylic acid provides the unsaturated (-C=C-) polymeric reaction site and a single ester group (Figures 3 and 5). The formal name of the vinyl ester is derived by the type of epoxy used. When starting with a bis A epoxy resin, the result is a bis A vinyl ester (Fig. 5) that has the toughness and improved wetting of the parent epoxy and the reactive polymerization sites of a polyester. From a molecular basis, bis A epoxy vinyl esters have better chemical resistance than polyesters because there are typically only two reaction sites and only two ester groups per molecule. In addition, all the ester groups are shielded with methyl groups (Fig. 5), thereby providing additional resistance to chemical degradation via hydrolysis.

Processing--With regard to processing, each resin has its own advantages or disadvantages, depending on the application. Polyesters and vinyl esters are diluted with styrene monomer which also participates in the free radical initiated polymerization. Free radicals for the cure are generated with the use of a small amount (i.e., 0.5-2.5% by weight) of a peroxide. Styrene also allows manufacturers to provide very low viscosity resins that easily wet out reinforcing fibers, fabric, and/or mineral fillers. Epoxy resins require a reactive equivalent of a curing agent (i.e., 10-115% by weight), which also strongly dictates cure times, the final physical properties, and chemical resistance. By far the most common curing agents are the family of modified amines. Epoxy resins are generally much more viscous and are sometimes difficult to wet-out fibers and fillers at ambient temperatures, but processing can be optimized with the use of reactive diluents and low viscosity curing agents.

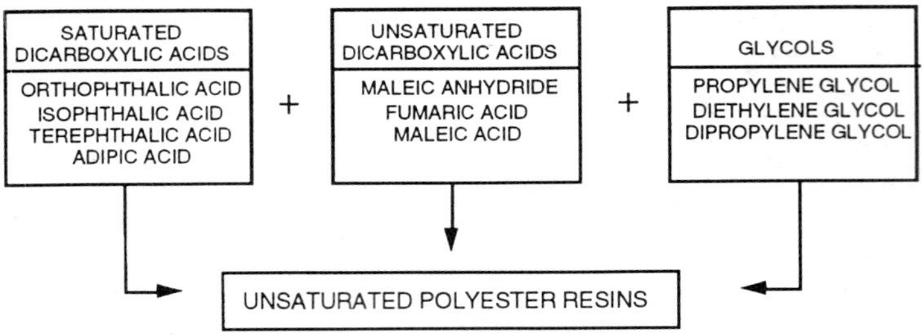

FIG. 1--Flow chart of common chemical components available for manufacturing different types of unsaturated polyester resins.

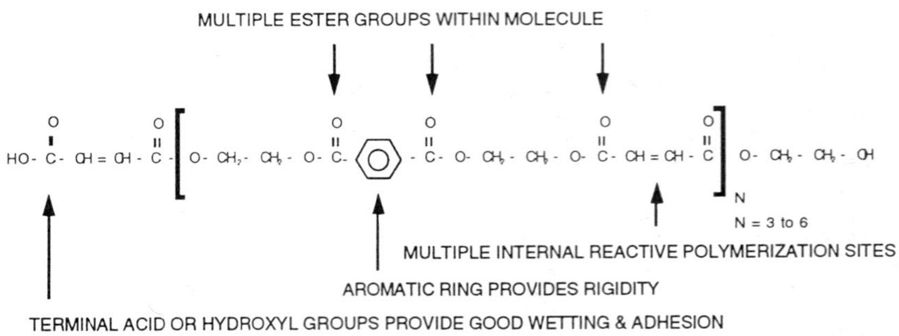

FIG. 2--Generic structure of an isophthalic polyester resin.

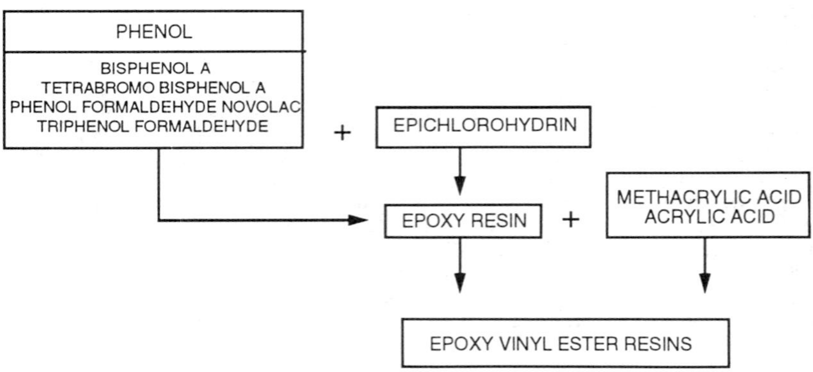

FIG. 3--Flow chart of common chemical components available to manufacture epoxy and epoxy vinyl ester thermosetting resins.

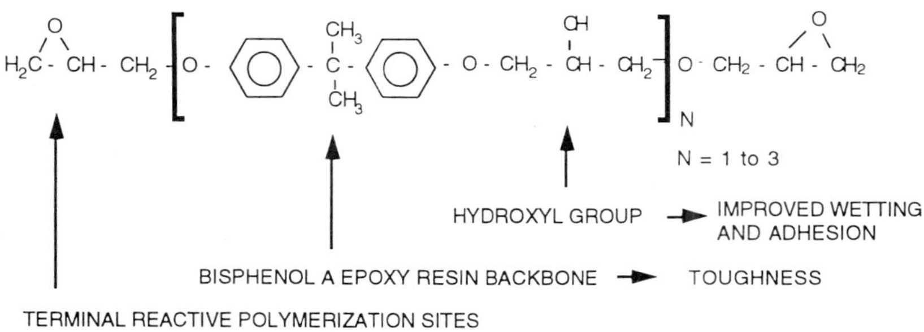

FIG. 4--Generic chemical structure of a bisphenol A epoxy resin.

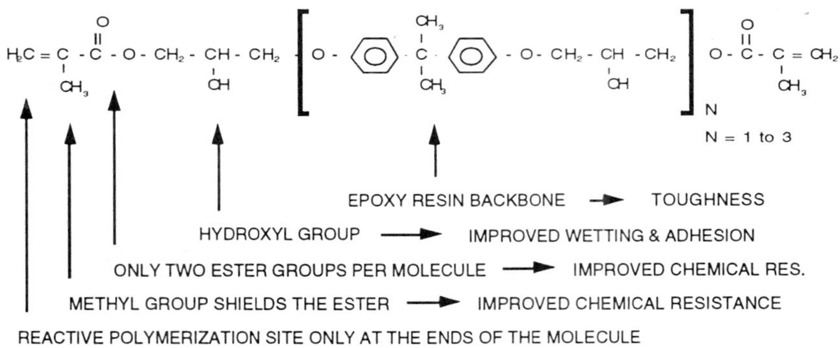

FIG. 5--Generic structure of a bisphenol A epoxy vinyl ester resin.

Short and Long Term Physical Properties

Introduction--As mentioned in the introduction of this paper, there is a distinct difference between short and long term physical properties, as well as the test methodologies used to distinguish the two. Through the use of examples and data, the distinction between the two and the type of information obtained will be clarified. Typically, short term properties are confirmatory or qualifying measures of minimum material properties obtained by relatively quick test methods. Flexural, tensile, and compression tests are examples of material evaluation techniques that rapidly load a specimen to failure and require minutes to perform. Short term properties are important for minimum requirements but are not necessarily an indicator of long term material performance. Long term properties such as creep (flexural, tensile), chemical resistance, and strain corrosion are are intended to ascertain the performance of a material or manufactured article exposed to lower level, or "real world" environmental stresses. When using long term testing protocols it is also desirable to test a material to failure, except failure should occur slowly over a long period of time. Unfortunately, long term testing can require years to obtain meaningful distinctions and is often accelerated to a reasonable degree to obtain meaningful comparisons in months instead of decades. Practical experience with the test methods, materials, knowledge of the actual service conditions, and common sense are a researcher's guide in determining reasonable accelerated test conditions that yield meaningful data within a lifetime. Only through carefully designed and controlled experiments can one make conclusions about long term material performance with the use of short term test data.

Flexural and Tensile--The flexural and tensile data given in Table 1 is that of neat resin properties not having any fillers or fibers added. The data in Table 2 was generated by testing CIPP resin/felt composites prepared as described in the Materials section of the paper. Both sets of data have been provided to better illustrate differences between neat resin and resin/felt composite properties, as well as providing a rudimentary basis by which one might estimate resin/felt composite properties from neat resin data that is more readily available. There are four key areas of interest in these two data sets that will be emphasized. First, there is not a tremendous difference between the flexural and tensile moduli of the neat resins (Table 1) and the CIPP resin/felt composites (Table 2). The polyester felt does not stiffen the composite and primarily functions to carry and hold the resin, as well as form the thickness and diameter of the tube. Therefore, the modulus for any neat resin would be expected to be similar to that of a CIPP resin/felt composite. Secondly, the flexural and tensile strengths of these CIPP resin/felt composites are less than the neat resins. The reduction of strength is a result of the short polyester felt fibers pulling apart from one another, breaking, and/or resin disbonding from the fibers. Thirdly, the flexural and tensile properties of the three resin classifications do have recognizable differences. Of these representative samples, the two Iso polyesters had high moduli and the lowest average strength and elongations compared to the other resins. Both bis A epoxy vinyl esters had a balance of high moduli and also had the highest average strengths and elongations at failure among the test group. The bis A epoxy resin/curing agent systems had quite different properties as a result of the different curing agents used to cure the epoxy resin. The ER-1 system had the lowest modulus and highest elongation in the test group. The ER-2 system was slightly stiffer, but had lower strength and flexibility than ER-1 and more or less behaved similarly to VER-1. In general, because of the toughness and flexibility imparted by the bis A molecular backbone (Figures 4 and 5) the epoxy and epoxy vinyl ester resins had the highest strengths and elongations among the test group. The multiple internal reaction sites and rigidity of the isophthalic acid in the Iso polyesters (Fig. 2) results in resins having high modulus, but comparatively lower strengths and elongations. Lastly, the data of all resin types given in Table 2 is somewhat higher than an average data set obtained from field samples used to qualify installed CIPP. One might typically expect flexural and tensile data from CIPP field samples to be in the range of 60-90% of that which can be obtained through careful preparation in the laboratory. Uncontrollable factors such as the equipment, geometric, and environmental conditions influence properties. Controllable factors such as properties of the chosen resin, cure system, cure schedule, and contractor experience also contribute to the variability of this process.

TABLE 1--Flexural (Flex.) and tensile (Tens.) modulus (Mod.) and Strength (Str.) of neat resin samples made with the given resins.

Test Prop.	VER-1	VER-2	ER-1	ER-2	PER-1	PER-2
Flex. Mod.	3.59 GPa (520 Kpsi)	3.52 GPa (510 Kpsi)	3.03 GPa (440 Kpsi)	3.24 GPa (470 Kpsi)	3.65 GPa (530 Kpsi)	3.24 GPa (470 Kpsi)
Flex. Str.	130.3 MPa (18.9 Kpsi)	130.3 MPa (18.9 Kpsi)	86.9 MPa (12.6 Kpsi)	107.6 MPa (15.6 Kpsi)	124.1 MPa (18 Kpsi)	117.2 MPa (17 Kpsi)
Tens. Mod.	3.45 GPa (500 Kpsi)	3.45 GPa (500 Kpsi)	2.69 GPa (390 Kpsi)	2.89 GPa (420 Kpsi)	3.59 GPa (520 Kpsi)	3.45 GPa (500 Kpsi)
Tens. Str.	63.4 MPa (9.2 Kpsi)	64.1 MPa (9.3 Kpsi)	48.9 MPa (7.1 Kpsi)	64.1 MPa (9.3 Kpsi)	60.0 MPa (8.7 Kpsi)	58.6 MPa (8.5 Kpsi)

TABLE 2--Flexural (Flex.) and tensile (Tens.) modulus (Mod.), strength (Str), and percent strain (Stn.) or elongation (Elon.) for resin/polyester felt CIPP composites made with the given resins.

Test Prop.	VER-1	VER-2	ER-1	ER-2	PER-1	PER-2
Flex. Mod.	3.45 GPa (500 Kpsi)	3.41 GPa (495 Kpsi)	2.72 GPa (394 Kpsi)	2.99 GPa (433 Kpsi)	3.54 GPa (513 Kpsi)	3.57 GPa (518 Kpsi)
Flex. Str.	76.5 MPa (11.1 Kpsi)	85.5 MPa (12.4 Kpsi)	73.8 MPa (10.7 Kpsi)	61.4 MPa (8.9 Kpsi)	65.5 MPa (9.5 Kpsi)	58.6 MPa (8.5 Kpsi)
Max. Stn.	2.5 %	3.0 %	3.5 %	2.2 %	2.0 %	1.7 %
Tens. Mod.	3.45 GPa (500 Kpsi)	3.17 GPa (460 Kpsi)	2.52 GPa (365 Kpsi)	2.65 GPa (384 Kpsi)	2.83 GPa (410 Kpsi)	... ...
Tens. Str.	44.8 MPa (6.5 Kpsi)	55.2 MPa (8.0 Kpsi)	48.9 MPa (7.1 Kpsi)	42.1 MPa (6.1 Kpsi)	35.2 MPa (5.1 Kpsi)	... ...
Tens. Elon.	1.8 %	2.6 %	2.8 %	1.8 %	1.5 %	...

Chemical Resistance--The acids, bases, and oxidizing agents selected for this study were chosen because they are naturally found in the sewer system or are introduced by homeowners, businesses, or sanitation departments [1,2,4,9]. This idealized testing protocol includes identical sample preparation and testing conditions in order to provide a good comparison of the relative chemical resistance of the six resins evaluated in a broad range of chemical effluents. The chemical resistance data in Figures 6-14 represent retention of flexural modulus versus time up to 12 months of exposure to the aforementioned chemicals, and the resins were abbreviated as previously noted in the Materials section. Coupon thickness and weight change information obtained at the 1, 3, 6, and 12 month evaluations is presented in Table 3. The data in Figures 7-14 is represented in log-log plots and statistically fit to represent retention of flexural modulus as a function of time. Statistically fit data such as this can then be used to make predictions about future performance by extrapolating forward in time [10]. The flexural moduli (Figures 7-14) and strength data (not graphically represented) were extrapolated to obtain estimates of the 25 year retention of these two properties (Table 4). Implicit with this data is the assumption that the specimens continue to perform in a consistent manner as they did in the first 12 months of the evaluation. However, there are several examples (Table 4) where the flexural properties of VER-1, VER-2, or ER-2 were greater than or equal to the control after 12 months' exposure to some of the chemicals tested. Therefore, 25 year flexural property retention predictions that would have been greater than 100% were given as 100+ in Table 4, since it unrealistic to predict a continued increase after long term chemical exposure.

A number of generalized trends are readily apparent from the data in Figures 6-14 and Tables 3 and 4. First, the retention of flexural modulus and strength closely correlate for all the resins in this experimental group (Table 4) [11,12]. The only exceptions to this generalization are the performance of ER-1, PER-1, and PER-2 in potassium permanganate, which will be addressed later. Secondly, the basic chemicals (i.e., sodium hydroxide and ammonium hydroxide) and the oxidizing agents (i.e., bleach and potassium permanganate) were in general more severe than the acids for all three classifications of resin. Thirdly, reduction of flexural properties correlated closely with large changes in test coupon thickness and weight change (Table 3). Lastly, the performance of both epoxy vinyl esters were extremely similar to one another, as was the performance of both Iso polyesters. However, the chemical resistance of the bis A epoxy resin was extremely dependent on the type of curing agent used to cure the resin.

In general, the PER-1 and PER-2 resins were most resistant to the acids, mildly susceptible to hydrogen peroxide, and very susceptible to bleach, potassium permanganate, sodium hydroxide, and ammonium hydroxide. Of the chemicals that most affected both Iso polyesters, potassium permanganate was the most subtle since testing through 12 months did not show severe loss of flexural properties (Fig. 8, Table 4). However, large increases in coupon thickness and weight gain (Table 3) indicate severe polymer swelling which typically leads to premature failure of the resin system. Among the chemicals tested the most severe is the sodium hypochlorite, which is the active ingredient most often used in household bleach, bathroom, and drain cleaners. Initially this protocol was started with 2.5% bleach (Fig. 7), but resin failure occurred so quickly that a second evaluation was started with 0.5% bleach (Fig. 6) so that the study would last a full 12 months. However, even the 0.5% bleach resulted to complete failure by 9 months.

Among the two bis A epoxies the multiamine/imidazole curing agent used in the ER-2 system is much more chemically resistant to all the chemicals tested than the flexibilized amine curing agent used to cure the ER-1 system. The ER-1 and ER-2 resin systems were both resistant to the basic chemicals (sodium hydroxide and ammonium hydroxide). The ER-1 system was mildly resistant to the acids, but extremely susceptible to the oxidizing agents (bleach, hydrogen peroxide, and potassium permanganate). Exposure of ER-1 to both concentrations of bleach and

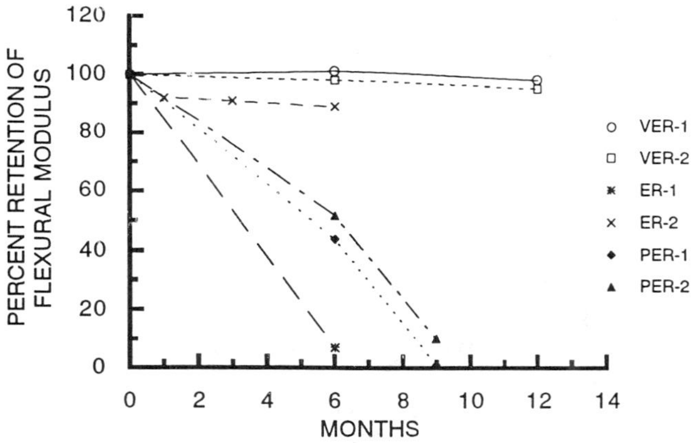

FIG. 6--Chemical resistance of CIPP to 0.5% sodium hypochlorite (bleach).

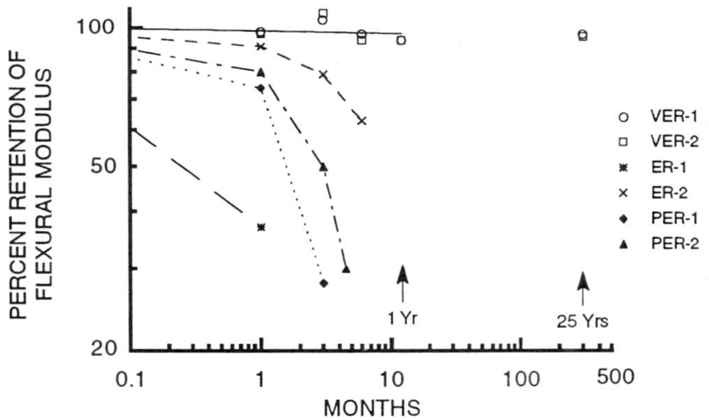

FIG. 7--Chemical resistance of CIPP to 2.5% sodium hypochlorite (bleach).

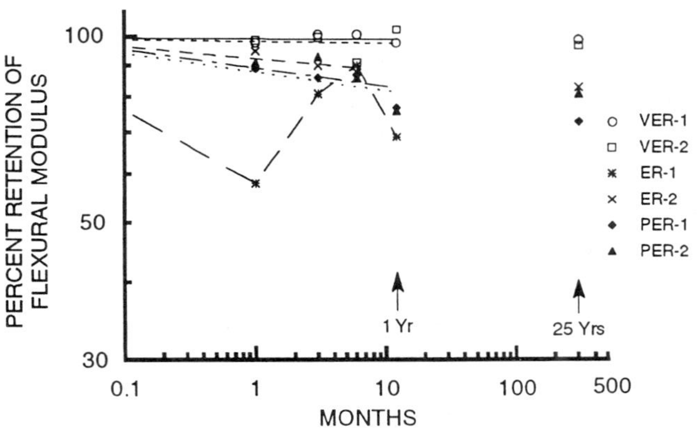

FIG. 8--Chemical resistance of CIPP to 5% potassium permanganate.

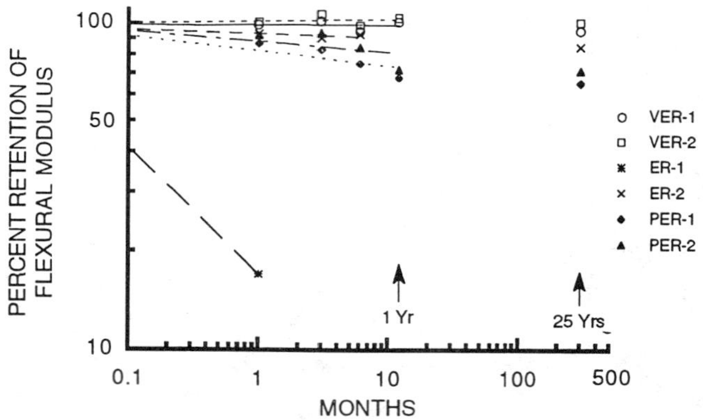

FIG. 9--Chemical resistance of CIPP to 5% hydrogen peroxide.

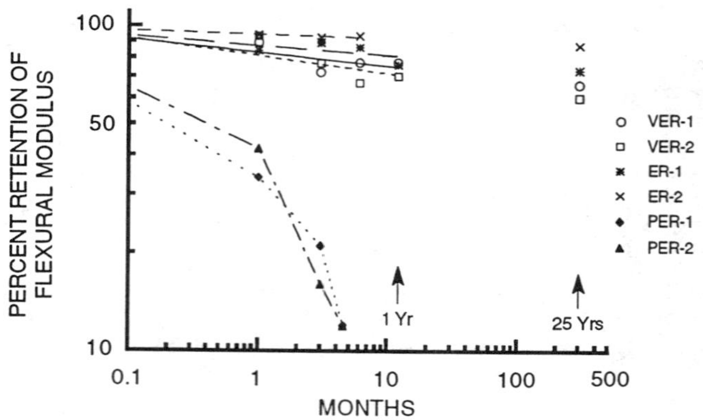

FIG. 10--Chemical resistance of CIPP to 5% sodium hydroxide.

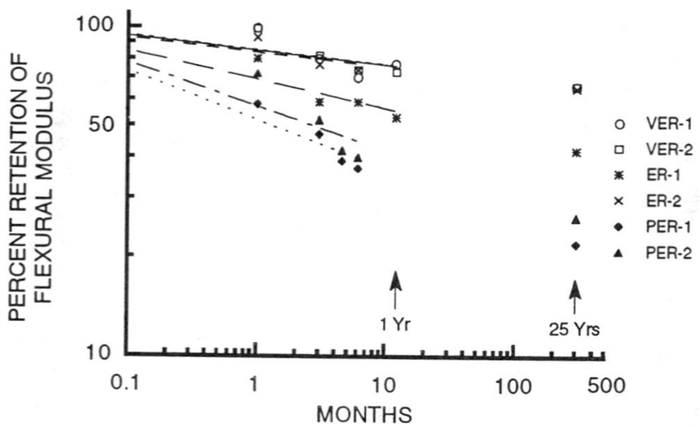

FIG. 11--Chemical resistance of CIPP to 5% ammonium hydroxide.

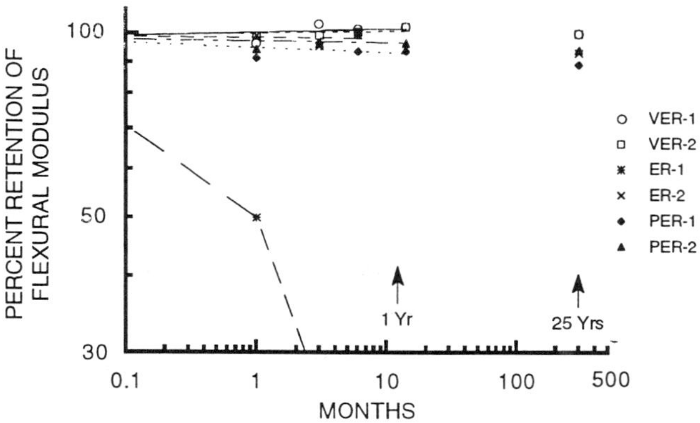

FIG. 12--Chemical resistance of CIPP to 25% sulfuric acid.

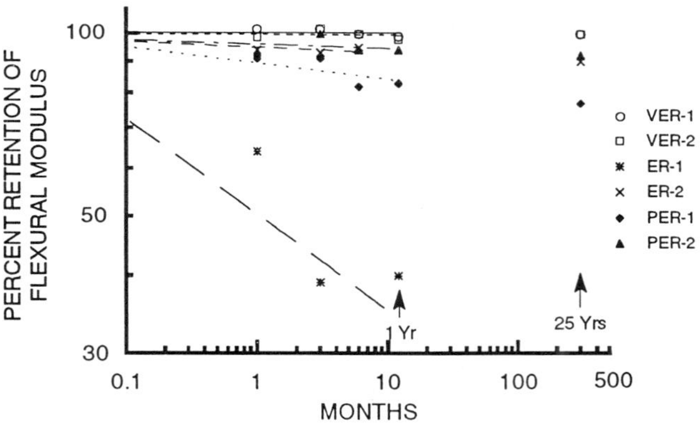

FIG. 13--Chemical resistance of CIPP to 20% hydrochloric acid.

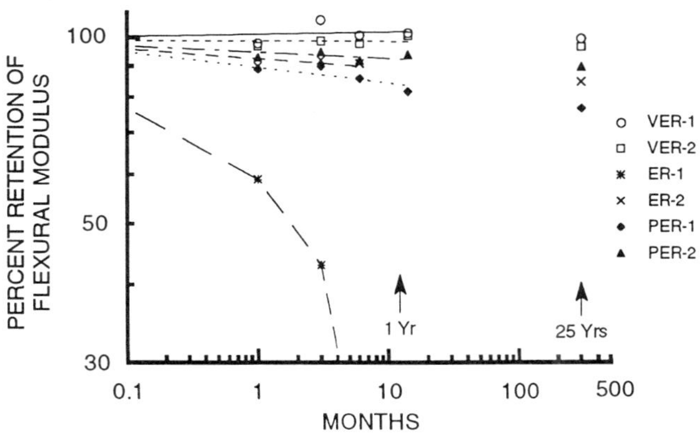

FIG. 14--Chemical resistance of CIPP to 5% nitric acid.

TABLE 3--Test coupon percent change of thickness and weight changes measured at 1, 3, 6, and 12 months (Mon.) during exposure to the given chemicals.

Chemical Tested	Property	Bis A Epoxy Vinyl Ester-1			Bis A Epoxy Vinyl Ester-2			Bis A Epoxy-1		
		3 Mon.	6 Mon.	12 Mon.	3 Mon.	6 Mon.	12 Mon.	1 Mon.	3 Mon.	6 Mon.
0.5% Sodium Hypochlorite	Coupon Thickness	-0.4	-0.2	0.0	0.0	-0.4	-0.4	...	-4.5	-6.1
	Coupon Weight	0.4	0.4	-0.5	0.4	0.5	-0.5	...	-25.2	-47.9
2.5% Sodium Hypochlorite	Coupon Thickness	0.0	-1.1	-1.1	0.0	-0.4	-1.1	0	-0.8	-11.1
	Coupon Weight	0.1	-0.9	-2.0	0.2	-1.1	-1.8	-18.8	-47.8	-84.7
5% Hydrogen Peroxide	Coupon Thickness	-0.4	0.0	-0.4	0.4	0.0	0.0	20.2	40.7	...
	Coupon Weight	0.6	0.6	0.5	0.5	0.5	0.5	18.0	44.5	...
								3 Mon.	6 Mon.	12 Mon.
5% Sodium Hydroxide	Coupon Thickness	0.4	-0.4	-0.8	0.4	0.0	0.4	1.7	0.4	1.7
	Coupon Weight	-3.7	-3.8	-5.7	-2.3	-4.8	-7.0	2.2	2.2	1.3
5% Potassium Permanganate	Coupon Thickness	0.8	0.8	1.8	0.0	0.4	2.2	16.6	27.8	29.8
	Coupon Weight	0.8	2.5	4.6	0.8	2.1	3.2	35.0	47.6	50.4
5% Ammonium Hydroxide	Coupon Thickness	0.8	0.0	0.8	0.4	0.0	0.4	2.1	2.1	9.4
	Coupon Weight	0.6	-2.7	-4.1	0.0	-3.1	-5.1	1.8	-1.1	-8.6
25% Sulfuric Acid	Coupon Thickness	0.4	0.4	0.0	0.4	-0.4	0.0	12.4	14.2	12.4
	Coupon Weight	0.3	0.3	0.3	0.2	0.2	0.2	25.5	30.9	32.1
20% Hydro-chloric Acid	Coupon Thickness	0.0	0.0	0.0	-0.4	0.0	0.0	6.7	8.9	6.7
	Coupon Weight	0.3	0.4	0.4	0.3	0.2	0.3	9.9	11.4	12.1
5% Nitric Acid	Coupon Thickness	0.0	0.0	0.0	-0.4	0.0	-0.4	4.9	7.9	25.9
	Coupon Weight	0.4	0.5	0.5	0.4	0.4	0.5	6.6	11.6	27.2

TABLE 3(Continued)--Test coupon thickness and weight changes measured at 1, 3, 6, and 12 months (Mon.) during chemical exposure to the given chemicals.

Chemical Tested	Property	Bis A Epoxy-2			Isophthalic Polyester-1			Isophthalic Polyester-2		
		1 Mon.	3 Mon.	6 Mon.	1 Mon.	3 Mon.	6 Mon.	1 Mon.	3 Mon.	6 Mon.
0.5% Sodium Hypochlorite	Coupon Thickness	0.4	-0.8	0.0	...	-0.7	0.4	...	0.0	-0.4
	Coupon Weight	1.2	0.9	-0.1	...	-5.4	-12.0	...	-4.1	-12.2
2.5% Sodium Hypochlorite	Coupon Thickness	0.0	0.0	-0.4	0.8	0.7	...	0.0	-0.8	...
	Coupon Weight	0.9	-1.9	-6.6	-0.8	-1.8	-78.0	-1.2	-6.5	-47.7
5% Sodium Hydroxide	Coupon Thickness	0.4	0.4	0.4	-0.7	-0.4	0.4	-0.7	-1.5	-1.5
	Coupon Weight	1.6	1.7	1.9	-4.6	-6.6	-7.5	-3.2	-7.9	-6.9
					3 Mon.	6 Mon.	12 Mon.	3 Mon.	6 Mon.	12 Mon.
5% Hydrogen Peroxide	Coupon Thickness	0.0	1.2	0.8	0.0	-0.4	-0.4	-0.4	0.4	-0.4
	Coupon Weight	2.3	3.3	3.7	0.8	0.3	-2.2	0.7	0.4	-1.4
5% Potassium Permanganate	Coupon Thickness	1.1	1.8	2.5	2.8	11.3	14.4	1.2	3.9	11.2
	Coupon Weight	3.4	4.3	4.7	4.1	22.8	28.1	2.0	6.9	18.7
5% Ammonium Hydroxide	Coupon Thickness	0.4	0.0	0.4	0.7	-1.4	...	1.1	-0.8	...
	Coupon Weight	1.9	1.9	-0.4	-3.4	-11.2	...	1.0	-7.6	...
25% Sulfuric Acid	Coupon Thickness	0.4	0.4	0.4	0.4	-0.8	0.4	0.0	-0.8	-0.4
	Coupon Weight	1.3	1.3	1.5	0.8	0.8	1.0	0.4	0.4	0.4
20% Hydro-chloric Acid	Coupon Thickness	0.4	0.4	0.7	0.7	0.4	0.4	0.4	-0.4	-0.4
	Coupon Weight	1.6	1.7	2.0	1.4	1.4	0.9	0.5	0.5	0.6
5% Nitric Acid	Coupon Thickness	0.7	0.7	0.7	0.0	0.0	0.4	-0.4	-0.8	0.0
	Coupon Weight	2.0	2.0	2.4	1.2	1.4	1.5	0.7	0.7	1.0

TABLE 4--Estimated percent retention of flexural properties[1] after 25 years (Yrs) of chemical exposure for resin/polyester felt CIPP composites made with the given resins.

Chemical Tested	Flexural Property	VER-1 25 Yrs	VER-2 25 Yrs	ER-1 25 Yrs	ER-2 25 yrs	PER-1 25 Yrs	PER-2 25 Yrs
2.5% Sodium Hypochlorite	Modulus	97	96	0	...	0	0
	Strength	99	86	0	...	0	0
5% Sodium Hydroxide	Modulus	66	60	73	88	0	0
	Strength	77	53	76	72	0	0
5% Hydrogen Peroxide	Modulus[2]	94	100+	0	84	65	71
	Strength	100	77	0	84	76	74
5% Potassium Permanganate	Modulus	99	97	...	83	73	81
	Strength	89	86	27	60	47	65
5% Ammonium Hydroxide	Modulus	66	66	42	65	22	26
	Strength	64	55	38	54	25	24
25% Sulfuric Acid	Modulus	100+	100+	0	93	89	94
	Strength	96	92	0	100+	88	94
20% Hydro-chloric Acid	Modulus	100+	99	21	89	77	92
	Strength	100+	91	31	95	70	95
5% Nitric Acid	Modulus	100+	97	0	85	77	90
	Strength	100+	94	0	82	72	93

[1]Flexural property retention estimates based on 12 month corrosion data.
[2](100+) indicates the curve fit would predict physical property retention greater than 100%

hydrogen peroxide resulted in complete destruction of the resin within a few months (Figures 6, 7, Table 3). Also, exposure to potassium permanganate was equally destructive to this resin. The flexural modulus initially dropped, then increased (Fig. 8), which is believed to be caused by the severe swelling a weight gain of the resin. The ER-2 system generally has excellent chemical resistance to the wide range of effluents tested, but shows some signs of susceptibility to the oxidizing agents through 6 months of testing. The importance of slight acceleration of the test protocol is evident for ER-2. From Fig. 6 and Table 3 ER-2 appears quite resistant to bleach. However, the results shown in Fig. 7 and the weight losses in Table 3 indicate a much greater susceptibility to bleach than originally might have been anticipated by evaluating the resin only at 0.5% bleach concentration.

Both bis A epoxy vinyl esters had excellent chemical resistance to the acids and the oxidizing agents, and good resistance to the bases tested. Of the chemicals tested the bis A epoxy vinyl esters were mildly susceptible to sodium and ammonium hydroxide (Fig. 10, 11 and Table 3) but were not significantly different from that of the epoxy resins. However, note that the bis A epoxy resins gain weight when exposed to caustic and the epoxy vinyl esters lose weight.

CONCLUSION

In summary, there were demonstrated differences in the physical properties and chemical resistance of the three classes of thermosetting resins evaluated. The flexural and tensile properties of these resins were considered to be good qualifying measures of the ultimate performance of each resin. Within this test group both Iso polyesters had high moduli, but low strength and strain. The epoxy vinyl esters had a balance of high moduli, strength, and strain. The bis A epoxy resins were very dependent on the curing agent, with ER-1 having the highest flexibility and ER-2 having properties similar to VER-1. The one year chemical resistance evaluations showed a number of common trends for the CIPP resin/felt composites. The percent reduction of flexural modulus generally correlates well with the percent reduction of flexural strength for each resin in each chemical chosen for testing. Low flexural property retention or failure correlated with relatively large weight and thickness changes, and weight change measurements are critical in interpreting more subtle changes. Evaluating two bis A epoxy vinyl esters and two Iso polyesters demonstrated how similar chemistry types appear to have similar performance features. However, the same bis A epoxy cured with a different curing agent significantly affects physical properties and chemical resistance.

Based on a 12 month chemical resistance study the Iso polyesters were good in acids, but failed in the basic chemicals, bleach, and potassium permanganate. The bis A epoxy resins are good in basic chemicals, but ranged from poor to excellent in the acids and oxidizing agents depending on the type of curing agent used. The bis A epoxy vinyl esters showed an ability to resist acids, bases, and oxidizing agents with little or no loss of physical properties during study.

Lastly, this study demonstrated the distinction between short and long term properties. Given the short term data (Tables 1 and 2), one could not have predicted the long term chemical resistance without more knowledge about the chemical composition of the individual resins and previously conducted chemical resistance screening. Although all the resins evaluated have acceptable short term physical properties, they are certainly not all acceptable for use in the full range of chemical agents tested.

REFERENCES

[1] "Hydrogen Sulfide Corrosion in Wastewater Collection and Treatment Systems," Report PB92-117985, Report to Congress, Sept. 1991.

[2] "Process Design Manual for Sulfide Control in Sanitary Sewerage Systems," Report EPA 625/1-74-005, U.S. EPA Technology Transfer, Oct. 1974.

[3] "Sewer System Infrastructure Analysis and Rehabilitation Handbook," Report CERI-91-42, U.S. EPA Office for Environmental Research Information, July 1991.

[4] "Technology Transfer Seminars: Sewer System Infrastructure Analysis and Rehabilitation," Report CERI-91-51, U.S. EPA Office for Environmental Research Information, July 1991.

[5] Parker, C.D., "Species of Bacteria Associated with the Corrosion of Concrete," Nature, Vol. 159, 1947, pp 439-40.

[6] Badia, Jaime, et. al., "Caustic Spray for Sewer Crown Corrosion Control", Proceedings of the ASCE International Conference on Pipeline Infrastructure, Aug. 1993.

[7] Marshall, G., Batis, G., "Preparing Collection Systems for Water Conservation", Water Environment & Technology, Vol. 5, No. 8, 1993, pp 52-57.

[8] Standard Specification for Public Works Construction (Green Book), BNi Building News, Los Angeles, 1991.

[9] Jeyapalan, J.K., "Prediction and Control of Sulfide Corrosion in Concrete Sewer Infrastructure and Rehabilitation", Corrosion Forms and Control for Infrastructure, STP 1137, Victor Chaker, Ed., American Society for Testing and Materials, Philadelphia, 1992, pp 273-283.

[10] "How Ingredients Influence Unsaturated Polyester Properties," Bulletin IP-70a, Amoco Chemical Company.

[11] Kleweno, D.G., "Characterization of Resins used for Cured-in-Place Pipe Rehabilitation", Proceedings of the North American NO-DIG '93 Conference, May 1993.

[12] Kleweno, D.G., "Chemical Resistance of Thermosetting Resins used for Cured-in-Place Pipe", Proceedings of the ASCE International Conference on Pipeline Infrastructure, Aug. 1993.

King H. Lo[1], and Jane Q. Zhang[1]

COLLAPSE RESISTANCE MODELING OF ENCASED PIPES

REFERENCE: Lo, K. H. and Zhang, J. Q., "**Collapse Resistance Modeling of Encased Pipes**," Buried Plastic Pipe Technology: 2nd Volume, ASTM STP 1222, Dave Eckstein, Ed., American Society for Testing and Materials, Philadelphia, 1994.

ABSTRACT: Two different collapse models are developed for predicting the collapse resistance of encased pipes for pipe rehabilitation applications. One model corresponds to an unsymmetrical collapse mode and the other one corresponds to a symmetrical collapse mode. Attention is focused here on modeling the effect of the small radial gap between the encased pipe and the host pipe on encased pipe collapse resistance. This small radial gap is assumed to have developed when installing the encased pipe into the host pipe. It increases in size when subjected to temperature change, swelling, and application of external hydrostatic pressure. The predictions by the models correlate well with some collapse test results on encased pipes.

KEYWORDS: encased pipe, collapse pressure, progressive buckling, radial gap effect

Plastic pipes and cured-in-place pipes are used more and more to rehabilitate buried pipes. For such applications, the ability of the plastic pipes and cured-in-place pipes (from here on will be referred to as the encased pipes) to resist collapse by external hydrostatic pressure and/or vacuum loading is an important design concern. Due to the constraint exerted by the host (rehabilitated) pipe, the collapse behavior of an encased pipe will be significantly different from that of an unsupported (stand-alone) pipe. In particular, the collapse resistance of an encased pipe can be many times higher than that of an unsupported pipe. This paper presents two simple engineering models for assessing the relative enhancement in collapse resistance that can be obtained with encased pipes. One model corresponds to an unsymmetrical collaspe mode and the other one corresponds to a symmetrical collapse mode.

Analytical modeling of the collapse behavior of radially constrained circular rings and pipes had been carried out by various researchers. Some representative examples of these modeling studies can be found in [1-6]. The general conclusion from these investigations is that the collapse resistance of an encased pipe is strongly influenced by the geometry of the initial imperfection in the pipe. The type of initial imperfection, used in all these studies, was in the form of an assumed crown deflection and an assumed detached length. The rest of the pipe was assumed, in the stress free state, to be in perfect contact with the confining medium.

[1]Shell Development Company, Westhollow Research Center, P. O. Box 1380, Houston, TX 77251-1380

In this study, no distinctive initial geometric imperfection is assumed to exist in the encased pipe. Instead, attention is focused here on modeling the effect of the small radial gap between the encased pipe and the host pipe on the collapse resistance. This small radial gap is assumed to have developed when installing the encased pipe into the host pipe. It can increase or decrease in size due to differential thermal expansion, swelling, and/or hydrostatic pressure trapped between the encased pipe and the host pipe.

The first part of the paper gives a brief description of the models. This is followed by correlation between model predictions and available test results.

COLLAPSE PRESSURE MODELS FOR ENCASED PIPES

The effect of the radial gap between the encased pipe and the host pipe is the key element in the development of the collapse models. Intuitively, a smaller radial gap would lead to a larger improvement in collapse resistance due to the stronger constraint imposed by the host pipe. Conversely, a larger radial gap would impose a weaker constraint on the encased pipe and thus would give rise to a smaller improvement in collapse resistance.

The problem considered here is modeled as a two dimensional plane strain buckling problem of the cross-section of the encased pipe. A schematic diagram of the encased pipe inside the host pipe is shown in Figure 1. The collapse process is modeled as a progressive buckling process. It starts with the encased pipe first coming into contact with the host pipe at one or two locations due to localized buckling of the encased pipe. As the hydrostatic pressure between the encased pipe and the host pipe increases, the contact area or buckled circumferential length increases. This process is initially stable since an increase in pressure is needed to induce buckling in the remaining unbuckled segment of the encased pipe. This process continues until no additional pressure increase can be taken and final collapse of the encased pipe has occurred. Model analysis starting with one contact location leads to an unsymmetrical collapse mode (Figure 2) and with two contact locations gives rise to a symmetrical collapse mode (Figure 3). In the following, the governing equations for the unsymmetrical collapse mode will be derived first to be followed by the equations for the symmetrical collapse mode.

Unsymmetrical Collapse Mode

To construct the progressive collapse model, the circumferential length of the encased pipe is assumed to remain unchanged during the buckling process. Further, we assume that once the buckled portion of the encased pipe comes into contact with the host pipe it will conform to the geometry of the host pipe while the remaining unbuckled segment of the encased pipe assumes the shape of an arc of a new circle. This new circle has an effective radius ρ and a new center O′ as shown in Figure 2(b). Note that the location of O′ will be different from that of the center of the circular cross-section of the host pipe. Assuming that the host pipe is rigid, the following geometric relations can be obtained for the unsymmetrical buckling mode.

$$2\pi R_2 - 2R_1(\pi-\beta) = 2\alpha\rho \tag{1}$$

$$\rho\sin\alpha = R_1\sin\beta \tag{2}$$

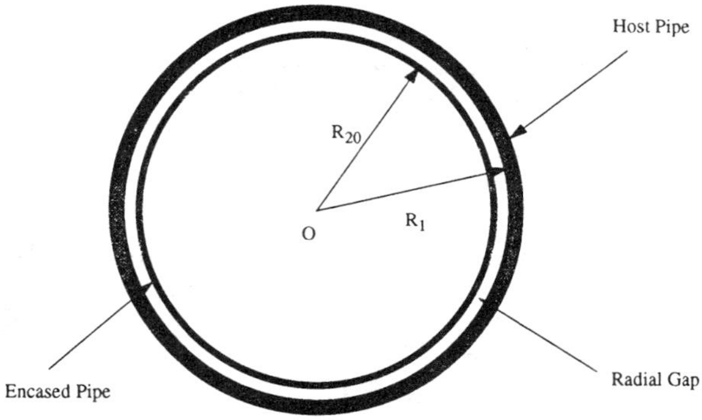

FIG. 1--A schematic diagram of encased pipe and host pipe.

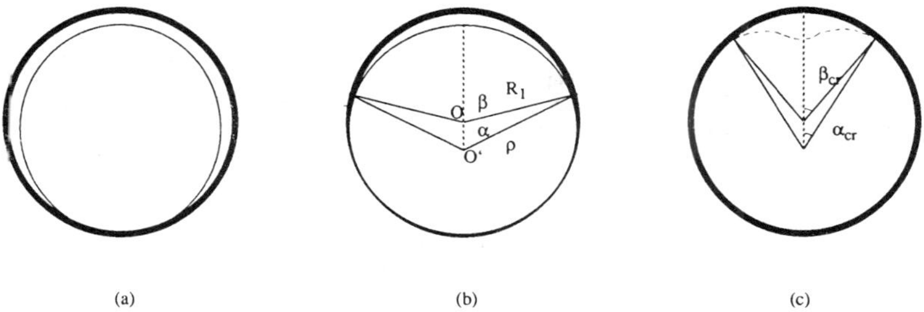

FIG. 2--Progressive buckling process for unsymmetrical collapse mode.

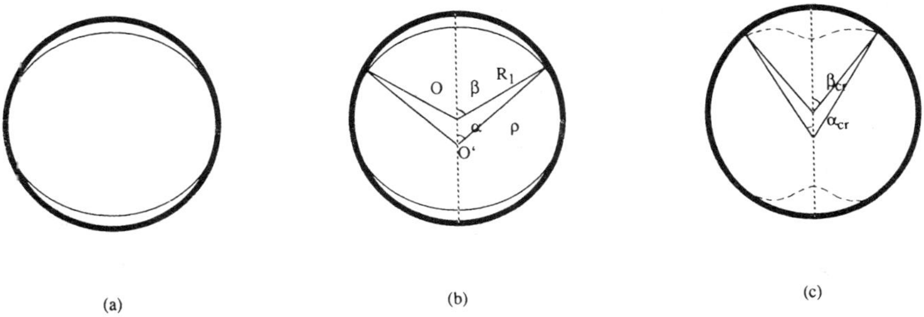

FIG. 3--Progressive buckling process for symmetrical mode.

where

R_1 = inside radius of the host pipe,
R_2 = current outside radius of the encased pipe (in the unbuckled state),
= $R_1 - \Delta$, and
Δ = current radial gap size between the encased pipe and host pipe.

In the above equations, α and β are the angular quantities defined in Figure 2(b). Here β is related to the portion of the encased pipe in contact with the host pipe and α is the angle sustained by the remaining unbuckled segment of the encased pipe.

Defining m as Δ/R_1, Eqs (1) and (2) can be rewritten in the following forms.

$$\frac{\beta}{\sin\beta} - \frac{\pi}{\sin\beta}m = \frac{\alpha}{\sin\alpha} \tag{3}$$

$$\rho = R_1 \frac{\sin\beta}{\sin\alpha} \tag{4}$$

Now, the unbuckled segment of the encased pipe (Figure 2(b)) can be viewed as a circular arch clamped at both ends and subjected to a uniformly distributed pressure p as shown in Figure 4. The critical pressure p that will induce buckling of the unbuckled segment of the encased pipe can be determined easily from the differential equation for the deflection of the circular arch and is given by [7]

$$p = \frac{EI}{(1-\nu^2)\left(\rho-\frac{t}{2}\right)^3}(k^2-1) \tag{5}$$

where

E = Young's modulus of the encased pipe,
ν = Poisson's ratio,
t = encased pipe wall thickness, and
I = moment of inertia of the pipe wall.

and k in the above equation is determined by satisfying the following transcendental equation.

$$k\tan\alpha = \tan k\alpha \tag{6}$$

Equations (3) - (6) form a set of equations for the determination of the collapse pressure for a given value of m. Starting with an assumed initial value for β, a set of values of α and ρ are determined from Eqs (3)-(4) and the corresponding values of k and p are obtained from Eqs (5)-(6). This process is then repeated for other smaller values of β. In general, the value of p increases as β decreases. The final collapse of the encased pipe is considered to have occurred when no further increase in the value of p can be obtained with smaller values of β. The corresponding value of p is then taken as the collapse pressure of the encased pipe.

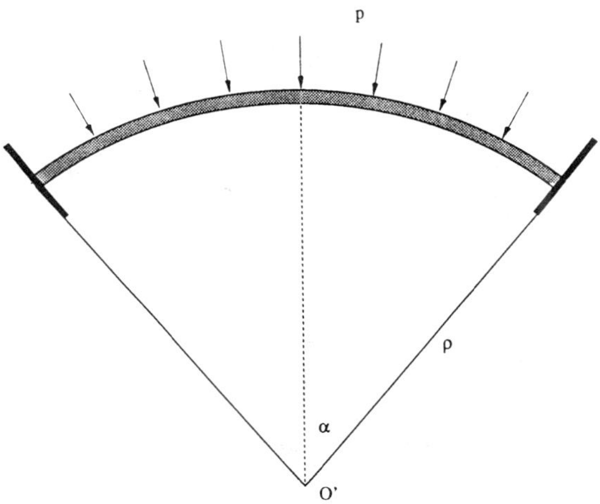

FIG. 4--A circular arch sujected to a uniformly distributed pressure.

It was mentioned earlier that the collapse resistance of an encased pipe can be many times higher than that of an unsupported pipe. This improvement in collapse resistance can be evaluated quantitatively by calculating the ratio of the collapse pressure of the encased pipe to that of the unsupported pipe. Calling this ratio the collapse pressure enhancement factor, H, and using the following equation for the collapse pressure of the unsupported pipe,

$$p=\frac{3EI}{(1-\nu^2)\,(R_{20}-\frac{t}{2})^3} \tag{7}$$

the enhancement factor, H, can be expressed as

$$H=\frac{(k^2-1)}{3}\left(\frac{R_{20}-\frac{t}{2}}{\rho-\frac{t}{2}}\right)^3 \tag{8}$$

where

R_{20} = initial outside radius of the encased pipe.

Using Eq (3), H can be rewritten as

$$H=\frac{k^2-1}{3}\left(\frac{\sin\alpha}{\sin\beta}\right)^3\left(\frac{R_{20}}{R_1}\right)^3\eta \tag{9}$$

where η is defined as

$$\eta = \left(\frac{1-n}{1-nq\left(\frac{\sin\alpha}{\sin\beta}\right)} \right)^3 \tag{10}$$

with $n = t/2R_{20}$ and $q = R_{20}/R_1$

For most pipe rehabilitation applications, the values of R_{20} and R_1 are sufficiently close to each other that the term $(R_{20}/R_1)^3$ is basically equal to unity, Similarly, the value of η can also be shown to be close to unity. Hence, the collapse pressure enhancement factor can be approximately expressed as

$$H = \frac{k^2-1}{3} \left(\frac{\sin\alpha}{\sin\beta} \right)^3 \tag{11}$$

It can be seen from the above equation that the enhancement H is more or less independent of the encased pipe and the host pipe geometry and is simply a function of the gap size ratio m.

For a given value of m, Eqs (3), (4), (6), and (9) form a set of equations for the determination of the enhancement factor H. The procedures to be followed are similar to those discussed earlier for the determination of the final collapse pressure of the encased pipe.

Symmetrical Collapse Mode

The equations for the symmetrical collapse mode can be derived following the same procedures as those for the unsymmetrical collapse mode. The collapse of the encased pipe is still modeled as a progressive buckling process (Figure 3). In this case, the encased pipe first comes into contact with the host pipe at two locations instead of one (Figure 3(a)). As the pressure between the encased pipe and the host pipe increases, the contact area or buckled circumferential length increases in a symmetrical manner (Figure 3(b)). The final collapse occurs as shown in Figure 3(c). Using the same assumptions as in the case of unsymmetrical buckling mode, the following geometric relations can be obtained for the symmetrical buckling mode.

$$2\pi R_2 - 2R_1(\pi - 2\beta) = 2\alpha\rho \tag{12}$$

$$\rho \sin\alpha = R_1 \sin\beta \tag{13}$$

Rewriting Eqs (12) and (13) using the definition of the gap size ratio m, we have

$$\frac{\beta}{\sin\beta} - \frac{\pi}{\sin\beta}\frac{m}{2} = \frac{\alpha}{\sin\alpha} \tag{14}$$

$$\rho = R_1 \frac{\sin\beta}{\sin\alpha} \tag{15}$$

It is important to point out that the only major difference between the symmetrical and unsymmetrical buckling modes is in the second term of Eqs (3) and (14). From this point onward, the balance of the formulation for the symmetrical buckling mode becomes identical to that of the unsymmetrical buckling mode. The final expression of the enhancement factor H for the symmetric buckling mode is the same as Eq (9) except that α, β, and $\underline{m}$ are now governed by Eq (14) instead of Eq (3).

Numerical Simulations

Using the equations developed for the unsymmetrical collapse mode, Figure 5 shows, as an illustrative example, the increase in the buckling pressure $\underline{p}$ as the buckling process progresses (or as the angle β decreases). Collapse of the encased pipe is assumed to have occured when no further increase in the buckling pressure is obtained with smaller values of β. The maximum buckling pressure is then taken as the collapse pressure of the encased pipe.

Figure 6 shows the predictions by the two models of the collapse pressure enhancement factor as a function of the ratio of the radial gap size to the host pipe inside radius. Here, the host pipe is assumed to be sufficiently rigid that no significant change in the host pipe inside diameter has occurred under the pressure in the gap. The enhancement factor in the figure is modified by the factor $(R_1/R_{20})^3$ to remove the geometric dependency of the results given in the figure. A value of η equals to one is used in calculating H.

It can be seen from Figure 6 that the enhancement factor decreases as the ratio of radial gap size to host pipe inside radius increases. For smaller gap size ratios, significant increase in collapse resistance is indicated. However, it is important to point out that such a large increase in collapse resistance can only be realized if the crush strength of the encased pipe is not exceeded. For gap size ratios larger than 5%, the enhancement factor remains more or less constant and has a value between 4 and 5.

Figure 7 shows the effect of the factor η on the enhancement factor. For encased pipes with standard dimension ratios (SDR = $2R_{20}/t$) between 40 - 100, the effect on the value of the enhancement factor by dropping η is less than 2.5 percent.

It is obvious that when the radial gap between the encased pipe and the host pipe becomes sufficiently large, the encased pipe would tend to behave more like an unsupported pipe. Hence, the predictions by the models will not be valid for very large gap size ratios.

Figure 8 shows the relationship between the enhancement factor H and β_{cr}. Here β_{cr} is one-half of the angle sustained by the remaining unbuckled segment of the encased pipe just before the final collapse of the encased pipe. It is interesting to note that for the same value of H, the values of β_{cr} for both the symmetrical and unsymmetrical buckling modes are almost identical. However, as can be seen from Figure 6, this corresponds to different radial gap size ratios for the two collapse modes.

CORRELATION WITH COLLAPSE TEST RESULTS

The predictions from the models on the collapse pressure enhancement factor are correlated with the test results given in [8]. Details of the collapse test program can be found in [8] and will not be given here. However, for the completeness of this paper, the pertinent test results are repeated below.

The first set of test data was obtained from encased pipe samples made with three different epoxy resin systems. These encased pipes were fabricated by direct inversion into steel pipes of 12 inches inside

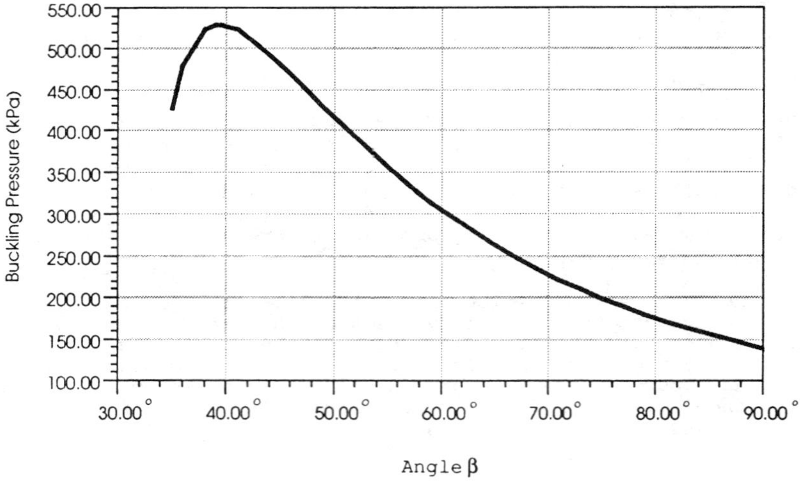

FIG. 5--Buckling pressure vs angle β.

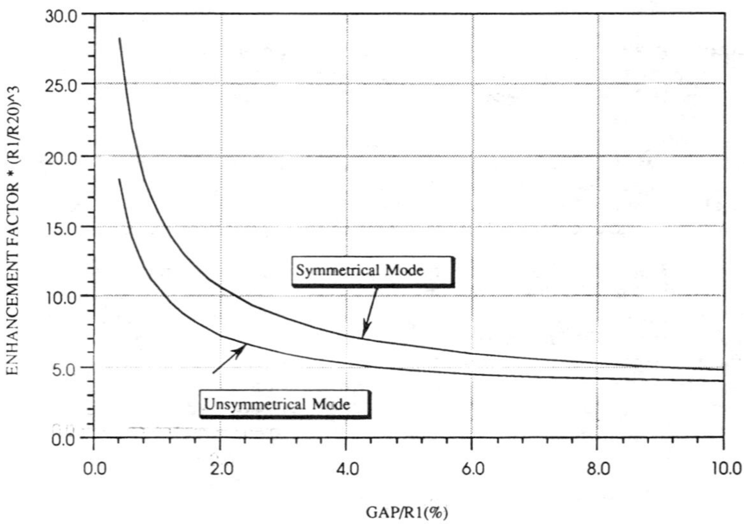

FIG. 6--Prediction of collapse pressure enhancement factor.

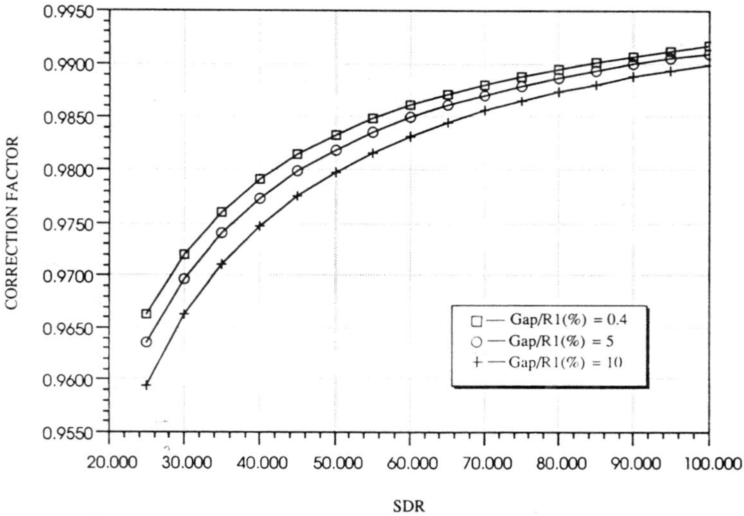

FIG. 7--Effect of η on collapse enhancement factor.

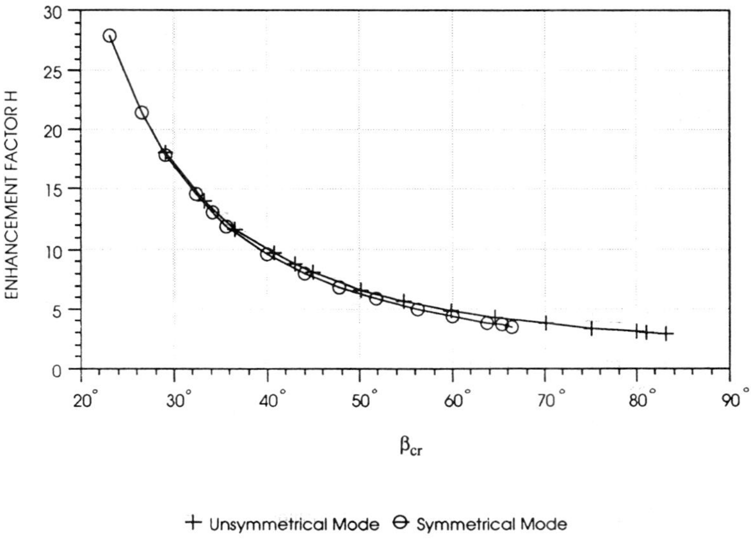

FIG. 8--Enhancement factor vs β_{cr}.

diameter. Table 1 shows the average stiffness and wall thickness of the encased pipe samples and Table 2 shows the collapse pressure test results. The collapse pressures for the corresponding unsupported pipes were calculated using Eq (7) and are shown in Table 3. The resulting collapse pressure enhancement factors were calculated and are shown in Table 4.

TABLE 1--Encased pipe modulus and wall thickness.

Resin - A		Resin - B		Resin - C	
average Modulus (GPa)	Thickness (mm)	average Modulus (GPa)	Thickness (mm)	average Modulus (GPa)	Thickness (mm)
2.551	5.545	3.227	5.689	3.406	5.514

TABLE 2--Collapse pressure of encased pipe.

Resin - A		Resin - B		Resin - C	
Sample #	Collapse pressure (kPa)	Sample #	Collapse pressure (kPa)	Sample #	Collapse pressure (kPa)
1-3	558.5	2-7	620.6	3-2	689.5
1-5	551.6	2-14	655.0	3-4	620.6
1-10	558.5	2-17	641.2	3-9	586.1
1-14	572.3	...	...	3-15	468.9
1-19	537.8	...	...	3-20	655.0
Average	555.7	Average	638.9	Average	604.0

TABLE 3--Collapse pressure of unsupported pipe.

Resin - A	Resin - B	Resin - C
Collapse Pressure (kPa)	Collapse Pressure (kPa)	Collapse Pressure (kPa)
36.96	50.61	48.54

To correlate the test results with predictions from the models, the size of the radial gap between the encased pipe samples and the steel pipes was calculated. This radial gap was developed upon cooling after curing in the encased pipe inversion process due to differential thermal expansion between the encased pipe and the steel pipe. Since the coefficient of thermal expansion of the encased pipe material is subtantially higher than that of the steel pipe, the initial size of the radial gap can be calculated as

$$\Delta_1=\alpha_0 R_1 \Delta T \tag{16}$$

where α_0 is the linear thermal expansion coefficient of the encased pipe material and ΔT is the difference between the glass transition temperature and the test temperature. Table 5 shows the values of α_0 and ΔT for the three different epoxy resin systems.

When hydrostatic pressure is introduced into the radial gap during the collapse test, the gap will increase in size under pressure. The amount of increase can be estimated by using the following equation [9]

$$\Delta_2=\frac{p}{E}(1-\nu^2)R_{20}\left(\frac{R_{20}^2+(R_{20}-t)^2}{R_{20}^2-(R_{20}-t)^2}-\frac{\nu}{1-\nu}\right) \tag{17}$$

The total gap size is equal to the sum of Δ_1 and Δ_2 and they are shown in Table 6.

Figure 9 shows the collapse test results superimposed on the predictions by the models. The correlation between the test results and the predictions is quite good. It is interesting to note that the test results fall within the prediction range of the two models. Since these two models are based on two different competing collapse modes, they can be used to explain the scatter in the test results obtained in the collapse test of encased pipes.

TABLE 4--Collapse enhancement factors of encased pipe.

Resin - A		Resin - B		Resin - C	
Sample #	Enhancement Factor	Sample #	Enhancement Factor	Sample #	Enhancement Factor
1-3	15.11	2-7	12.26	3-2	14.2
1-5	14.92	2-14	12.94	3-4	12.78
1-10	15.11	2-17	12.67	3-9	12.07
1-14	15.48	...	...	3-15	9.66
1-19	14.55	...	...	3-20	13.49
Average	15.03	Average	12.62	Average	12.44

TABLE 5--Thermal expansion coefficient α_0 and temperature difference ΔT.

Material	α_0	ΔT (°C)
Resin - A	9×10^{-5}/°C	45.0
Resin - B	7.2×10^{-5}/°C	88.89
Resin - C	7.2×10^{-5}/°C	100.0

Figure 10 shows the correlation between the predictions by the models and the collapse pressure test results provided by [10]. In this case, the initial radial gap size (before the application of external hydrostatic pressure) was estimated from measurement of the encased pipe and the steel pipe diameters. The final gap size was obtained by adding the contribution from external hydrostatic pressure calculated using Eq (17). It can be seen from the figure that good correlation is again obtained with the test results over a wide range of gap size ratios. In addition, the test results clearly demonstrate that as the gap size ratio increases, the enhancement in the collapse pressure resistance decreases. However, it is important to point out that a large initial gap size seldom exists in pipe rehabilitation. The test results shown in Figure 10 with large gap size ratios are only provided to clearly demonstrate the effect of gap size ratio on collapse resistance of encased pipes.

TABLE 6--Radial gap between the encased pipe and host pipe.

Sample #	Δ_1 (mm)	Δ_2 (mm)	$(\Delta_1+\Delta_2)/R_1$ (%)
1-3	0.6172	0.7696	0.909
1-5	0.6172	0.7595	0.903
1-10	0.6172	0.7696	0.909
1-14	0.6172	0.7874	0.921
1-19	0.6172	0.7391	0.890
2-7	0.9754	0.6553	1.069
2-14	0.9754	0.6934	1.093
2-17	0.9754	0.6782	1.083
3-2	1.0973	0.7137	1.186
3-4	1.0973	0.6426	1.139
3-9	1.0973	0.6071	1.116
3-15	1.0973	0.4851	1.037
3-20	1.0973	0.6782	1.162

DISCUSSION

The two models developed in this study are aimed at developing a simple method to determine the collapse resistance of encased pipes. Even though reasonable correlation has been obtained between predictions by the models and some available test results, more test data are needed to fully establish the validity of the models. Despite this, the effect of radial gap size on the collapse resistance of encased pipe is clearly demonstrated. Improvement to the models and/or new models are expected as further knowledge and understanding on encased pipe collapse resistance are accumulated.

As expected, the collapse pressure enhancement factor associated with the symmetric collapse mode is higher than that associated with the unsymmetrical collapse mode. At small gap size ratios, the difference in the predictions by the models is relatively large. At larger gap size ratios, the difference between the predictions diminishes. Although the observable final collapse mode of almost all of the encased pipes tested in the laboratory shows an unsymmetrical collapse mechanism, the tendency to develop symmetric collapse mode before final collapse occurs has been observed in some collapse tests. Hence, the collapse pressure measured in the collapse test might not be associated with only the unsymmetrical collapse mode. It could also be a reflection of the combined effect of these two competing collapse mechanisms. This might help to explain the scatter in the test data observed in the collapse test. The predictions on enhancement factor given in Figure 6 can, therefore, be regarded as the bounding values for the collapse pressure.

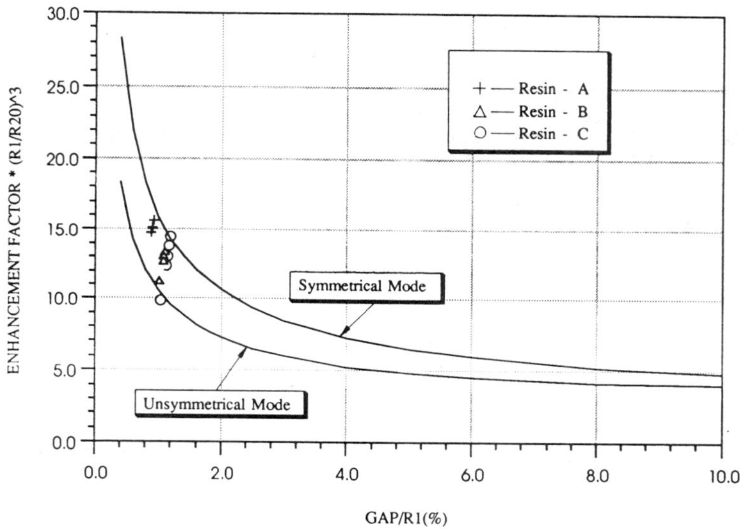

FIG. 9--Model prediction correlates with test data.

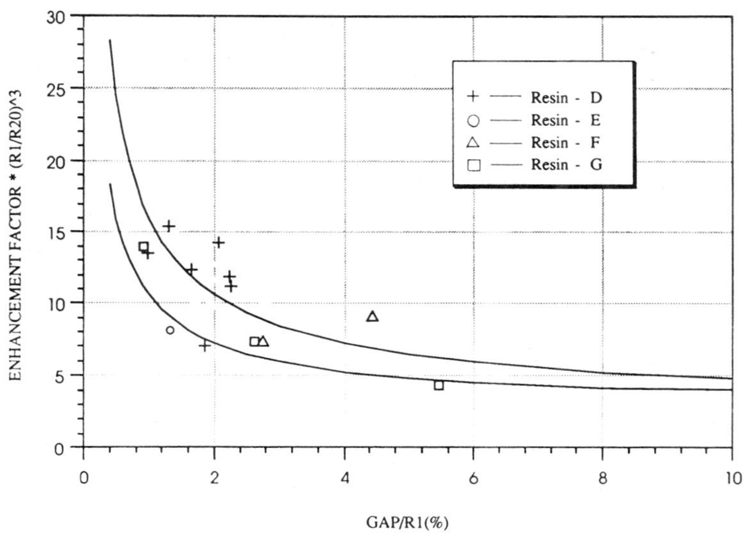

FIG. 10--Model prediction correlates with test data.

In addition to the short term collapse resistance, the long term collapse resistance of encased pipes is also an important design concern. When subjected to sustained hydrostatic pressure, the radial gap between the encased pipe and the host pipe will increase in size with time due to creep in the encased pipe. Hence, according to the models developed here, the enhancement factor and the corresponding collapse pressure will decrease with time due to creep. This prediction is qualitatively consistent with experimental observations. However, further theoretical and experimental work are needed to establish the utility of the models for assessing the encased pipe's long term collapse resistance.

REFERENCES

[1] Kyriakides, S. and Youn, S.-K., "On The Collapse of Circular Confined Rings Under External Pressure," International Journal of Solids Strutures, Vol. 20, No. 7, 1984, pp 669-713.

[2] Kyriakides, S. and Li, F.-S., "On The Response and Stability of Two Concentric, Contacting Rings Under External Pressure," International Journal of Solids Structures, Vol. 27, No. 1, 1991, pp 1-14.

[3] El-Bayoumy, L., "Buckling of a Circular Elastic Ring Confined to a Uniformly Contacting Circular Boundary," ASME Journal of Applied Mechanics, 1972, pp 758-766.

[4] Pian, T. H. H. and Bucciarelli, L. L., "Buckling of Radially Constrained Circular Ring Under Distributed Loading," International Journal of Solid Structures, Vol. 3, 1967, pp 715-730.

[5] Bucciarelli, L. L. and Pian, T. H. H., "Effect of Initial Imperfections on the Instability of a Ring Confined in an Imperfect Rigid Boundary," ASME Journal of Applied Mechanics, 1967, pp 979-984.

[6] Zagustin, E. A. and Herrmann, G., "Stability of an Elastic Ring in a Rigid Cavity," ASME Journal of Applied Mechanics, 1967, pp 263-270.

[7] Timoshenko, S. P.and Gere, J. M., "Theory of Elastic Stability," McGraw-Hill, New York, 1961.

[8] Lo, K. H., Chang, B. T. A., Zhang, Q., and Wright, W. J., "Collapse Resistance of Cured-In-Place Pipes," North American Society for Trenchless Technology, No-Dig '93, May 2-5, 1993.

[9] Venkatraman, B., and Patel, S. A., "Structural Mechanics with Introduction to Elasticity and Plasticity," McGraw-Hill, New York, 1970.

[10] Private Communication with Insituform Technologies, Inc.

Laboratory Testing

David W. Woods[1] and Steven R. Ferry[1]

COMPRESSIVE BUCKLING OF HOLLOW CYLINDERS: IMPLICATIONS FOR PRESSURE TESTING OF PLASTIC PIPE

REFERENCE: Woods, D. W. and Ferry, S. R., **"Compressive Buckling of Hollow Cylinders: Implications for Pressure Testing of Plastic Pipe,"** Buried Plastic Pipe Technology: 2nd Volume, ASTM STP 1222, Dave Eckstein, Ed., American Society for Testing and Materials, Philadelphia, 1994.

ABSTRACT: Compressive buckling of hollow cylinders due to internal pressure is a little-known phenomenon that can affect pressure testing of plastic pipe. High strength pipe materials with large dimension ratios particularly exhibit the phenomenon, but it can also occur in pipe with low dimension ratios (relatively thick walls) when the material is highly elastic or plastic at the test conditions. To successfully test such materials may require special constraints on test geometry and end closures. Specific examples of the phenomenon observed in the laboratory are discussed along with an analytical treatment from the literature.

KEYWORDS: plastic pipe, buckling, pressure testing, internal pressure

INTRODUCTION

The fact that plastic pipe can experience buckling is widely known and amply documented. Consult any handbook or design reference book in Mechanical, Civil, or Chemical Engineering and you will likely find analytical models and examples of two types of buckling that plastic pipe can experience. These are: column-type buckling under axial compressive loading of the wall of a hollow cylinder, and buckling under uniform radial compressive loading due to external hydrostatic pressure.

These types of buckling are familiar to most engineers. Technically trained people generally find these phenomena and their analysis to be straightforward and consistent with intuition. They are discussed here merely to contrast them with a third mode of buckling that can occur: compressive buckling due solely to internal hydrostatic pressure. This third buckling mechanism is rarely discussed in handbooks, and often strikes technically-trained persons who encounter it for the first time as counter-intuitive. The authors confess it struck them that way.

[1]Section Manager and Engineer II, respectively, Engineering Services, Hauser Laboratories, 2100 Central Avenue, Suite 100, Boulder, CO 80301.

THREE WAYS HOLLOW CYLINDERS BUCKLE

Column-Type (Euler) Buckling Under Axial Compression

The basic theory of how columns of any cross section buckle under axial compressive loads originates primarily with Euler [1], who showed that the critical buckling load is unrelated to material strength (yield or ultimate strengths), but is governed only by material modulus of elasticity and column geometry. Three formulas are commonly used to calculate critical buckling loads for the three cases that frequently occur in mechanical and structural design, as shown in Figure 1. In increasing magnitude, these are:

One end fixed; one end free -- This geometry is shown in Figure 1a, with the critical buckling load calculated using the following equation in any consistent system of units [2]:

$$F_{CR} = \pi^2 EI/4L^2 \tag{1}$$

where

F_{CR} = critical load or force required to cause buckling,

E = material modulus of elasticity,

I = moment of inertia,

L = column length.

Both ends pinned -- In this case the critical force becomes:

$$F_{CR} = \pi^2 EI/L^2. \tag{2}$$

Both ends fixed -- Fixing both ends raises the critical buckling force to:

$$F_{CR} = 4\pi^2 EI/L^2 \tag{3}$$

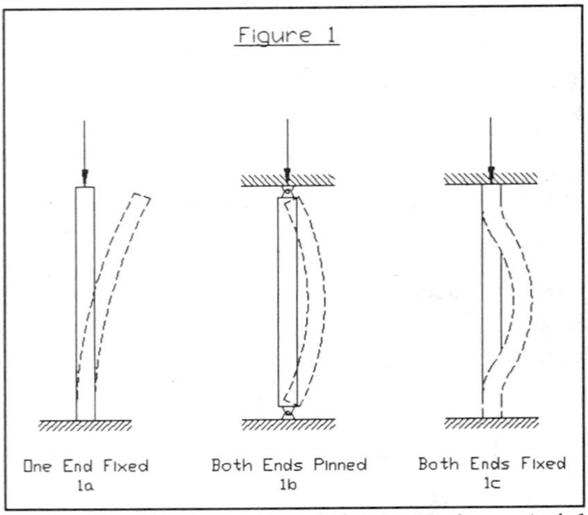

Figure 1 Column Buckling Under Axial Compressive Loading

These equations are typically modified to handle plastic materials by substituting a more relevant property for the elastic modulus, such as a tangent modulus at load, a reduced modulus, or a time-dependent modulus [3].

Tube Buckling Under External Hydrostatic Pressure

The collapse of thin-walled tubes under internal vacuum or external hydrostatic pressure is the second familiar type of buckling that hollow cylinders exhibit. It is something the designers of both piping systems and submarines must avoid.

The limit for this type of buckling is frequently expressed as a critical pressure, and can be calculated as [4]:

$$P_{CR} = [2E(1-\nu^2)]*(t/D)^3 \tag{4}$$

where

P_{CR} = critical buckling pressure

ν = Poisson's ratio

t = wall thickness

D = mean diameter

As with column-buckling, for plastic materials a long-term or time-dependent modulus is typically used for analysis of long-term performance. Because flexible pipe is not installed as a perfect cylinder, an ovality correction factor of the form

$$(r_o/r_i)^3 \quad (5)$$

is used, where

r_o = radius if perfectly cylindrical, and

r_i = major radius of curvature as installed.

Buckling Under **Internal** Hydraulic Pressure

The least familiar form of buckling that hollow cylinders exhibit is Euler-type buckling due to internal pressure. Designers of long hydraulic cylinders routinely concern themselves with the phenomenon, but it is seldom encountered in other fields.

Figure 2 illustrates buckling of a hydraulic cylinder under internal pressure. Although there is no direct axial loading of the cylinder walls, the cylinder/fluid system responds as an Euler column with two pinned ends, similar to Figure 1b. The total force on the piston rods is related to the internal cylinder pressure as

$$F = \pi r^2 P \quad (6)$$

Combining equation (6) with equation (2) and substituting $\pi r^3 t$ as the moment of inertia of a thin-walled cylinder yields a formula for critical internal pressure for buckling [5]:

$$P_{CR} = \pi^2 Etr/L^2 \quad (7)$$

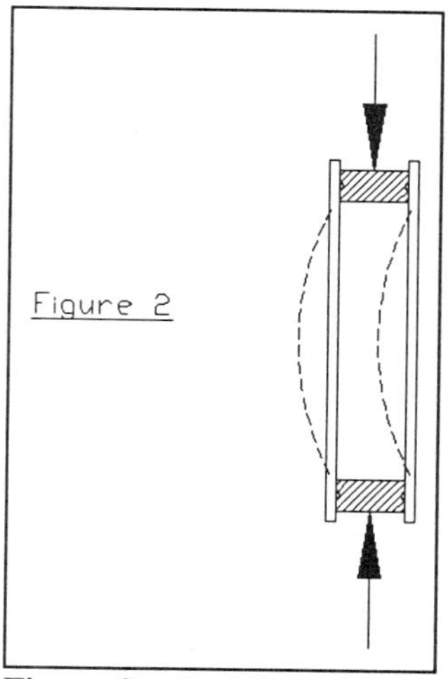

Figure 2 Buckling Under Internal Pressure

IMPLICATIONS FOR PRESSURE TESTING OF PLASTIC PIPE

ASTM test methods for hydrostatic pressure testing of plastic pipe and the specifications that reference the pressure test methods appear not to have considered buckling due to internal pressure in setting dimensional and equipment requirements. In certain instances, allowable dimensional and equipment combinations may need to be avoided in order to avoid buckling as the end point of the test.

ASTM Standard Test Method for Time-to-Failure of Plastic Pipe Under Constant Internal Pressure (D1589) and ASTM Standard Test Method for Short-Time Hydraulic Failure Pressure of Plastic Pipe, Tubing, and Fittings (D1599) both allow use of either free or restrained end-closures. In this context, a free end-closure fastens to the pipe so that internal pressure exerts axial force on the specimen, while a restrained end-closure can slide axially along the pipe, and relies on an internal or external structure to resist the end thrust from pressurization. Both methods also include requirements for the minimum length pipe specimen that can be tested, but specify no maximum length for a specimen.

When restrained end-closures are used to hydrostatically test a pipe specimen, the system becomes similar to the hydraulic cylinder example of Figure 2 and equation (7). The "pistons" become the restrained end-closures, and the axial force originates not from piston displacement, but from pumping fluid into the specimen. There are no axial forces acting directly on the pipe walls, yet the pipe/fluid system will buckle as an Euler column. Thus, by equation (7), if a sufficiently long pipe specimen is tested per D1598 or D1599 using restrained end-closures, the failure will occur not by bursting or ballooning, but by buckling.

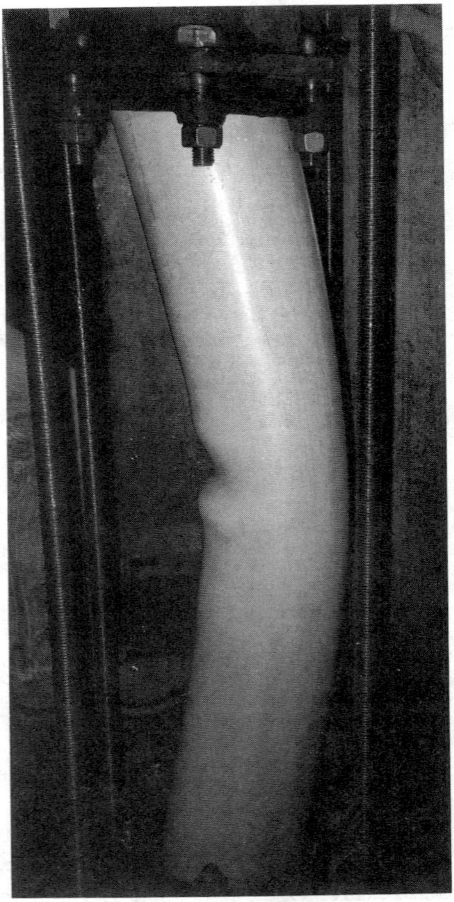

Figure 3 6 in IPS D1599 tested with restrained caps failed by buckling.

SPECIFIC EXAMPLES

We at Hauser Laboratories first encountered this phenomenon while conducting accelerated-regression Hydrostatic Design Basis (HDB) testing on a client's unique product. The pipe we were testing was 6 inch IPS nominal diameter with a Dimension Ratio (DR) of about 33 and a hoop tensile strength of over 75 MPa (11 000 psi). Thus, compared to most commercial plastic pipe, the product we tested had significantly higher strength and significantly thinner walls. The moment of inertia against buckling was lower relative to the burst strength of the pipe than for most commercial pipe.

When we attempted to perform a 23°C D1599 burst test on a 1 m specimen of this pipe, with 0.85 m of pipe exposed between restrained end-closures (slightly more than 5 diameters, as required by D1599), the pipe failed by buckling at a pressure that gave a hoop stress of about 78.5 MPa (11 400 psi), as shown in Figure 3. A subsequent test per D1598 with a hoop stress of about 66.2 MPa (9600 psi) produced a buckling failure in slightly less than 1 hr.

We then changed to free end-closures and continued to perform D1598 tests on additional specimens of the same pipe. These tests consistently resulted in classic hoop-tensile burst failures, as shown in Figure 4.

Recently, we observed the onset of buckling with a more unlikely geometry: 8 in IPS nominal pipe with a DR of 11 and an exposed length of 0.9 m. This geometry would appear to be completely robust against buckling, but after 1 000 hours sustained internal pressure creating 5 MPa (725 psi) hoop stress at a temperature of 80°C, each of the three specimens exhibited pronounced axial bowing. Had this sustained pressure test not been terminated at 1 000 hours, failure may ultimately have occurred either by bursting or by buckling: we cannot predict this from the data at hand. But the results clearly indicated that when the tested pipe is highly plastic at the test conditions, buckling may initiate even when the test geometry would appear to rule it out.

Figure 4 6 in IPS D1598 specimen tested with free caps failed by bursting.

CONCLUSIONS

Buckling of pipe due to internal hydrostatic pressure can occur whenever the internal pressure does not produce axial tensile loads on the pipe. The fact that ASTM pipe pressure test methods specify minimum, but not maximum, pipe lengths between end closures, while allowing the use of restrained end-closures, means that buckling can occur in certain tests. The propensity for buckling increases with increasing material strength, increasing plasticity, decreasing material modulus, and decreasing moment of inertia (increasing DR). The propensity to buckle is proportional to the test length squared.

REFERENCES

[1] Timoshenko, S., History of Strength of Materials, McGraw-Hill Book Co., New York (1953).

[2] Faupel, J. H., Engineering Design, pp 496-498, John Wiley and Sons, Inc., New York (1964).

[3] Timoshenko, S., Strength of Materials, Part II, D. Van Nostrand Co, Princeton, NJ, 3rd Ed. (1956).

[4] Mruk, S. A., "Thermoplastic Piping" p. D44, in Piping Handbook, 6th ed., edited by Nayyar, M. L., McGraw-Hill, Inc. (1992).

[5] [2], p. 537.

Ernest T. Selig[1], Leonard C. DiFrancesco[2] and Timothy J. McGrath[3]

LABORATORY TEST OF BURIED PIPE IN HOOP COMPRESSION

REFERENCE: Selig, Ernest T., DiFrancesco, Leonard C., and McGrath, Timothy J., "**Laboratory Test of Buried Pipe in Hoop Compression,**" Buried Plastic Pipe Technology: 2nd Volume, ASTM STP 1222, Dave Eckstein, Ed., American Society for Testing and Materials, Philadelphia, 1994.

ABSTRACT: A new test has been developed to study the behavior of buried pipe under hoop compression loading. The apparatus consists of a cylindrical steel vessel lined with an inflatable bladder. A length of pipe is installed at the center of the vessel and the annulus between the pipe and the bladder is filled with tamped sand. The test is conducted by incrementally increasing the bladder pressure while monitoring the pipe performance. Radial pressures of 380 kPa (55 psi) and diametral displacement of 37 mm (1.5 in.) have been obtained with corrugated pipes of 610 mm (24 in.) diameter. The technique is believed applicable to any size pipe with any pressure level. The test has demonstrated the significant circumferential shortening that can occur for plastic pipe sections with corrugated cross-sections. This produces beneficial positive arching in service. The test also provides a basis for determining plastic pipe wall design limits in compression.

KEY WORDS: buried pipe, plastic pipe, hoop compression, design limits, pipe test, laboratory test

Buried plastic pipes are subjected to earth pressures that produce both hoop compression stresses and ring bending stresses. ASTM Standard Test Method for Determination of External Loading Characteristics of Plastic Pipe by Parallel-Plate Loading (D2412) is used to evaluate ring bending stiffness and strength, but no test is available to evaluate performance in hoop compression, even though AASHTO [1] requires a design evaluation for local buckling of corrugations. Also, research has shown that, without internal pressure, corrugated pipe sections buried with substantial heights of cover may have hoop compression stresses greater than the ring bending stresses, and thus have no net tension stress in the pipe wall due to bending. A design limit based on wall compression, is therefore appropriate and so a test to determine

[1]Professor of Civil Engineering, University of Massachusetts, Amherst, MA 01003.

[2]Graduate Student, Department of Civil Engineering, University of Massachusetts, Amherst, MA 01003.

[3]Senior Associate, Simpson, Gumpertz and Heger, Inc., Arlington, MA 02174, and Doctoral Student, University of Massachusetts, Amherst, MA 01003.

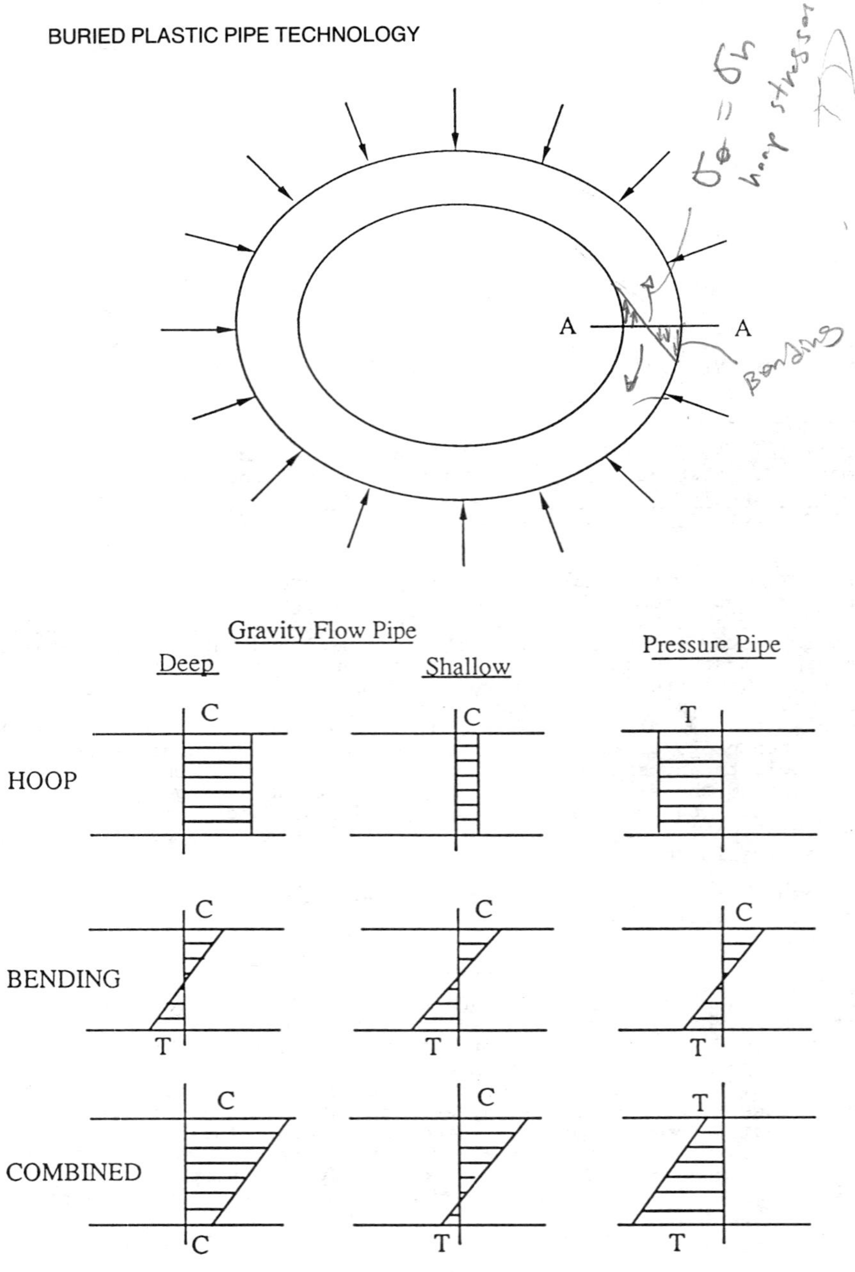

Fig. 1 Combined Hoop and Bending Stresses at A-A

this limit is required. This paper describes a new laboratory test for evaluating the behavior of plastic pipe in hoop compression with a loading that closely models the actual field loading.

TEST CONCEPT

The general nature of the combined hoop and bending stresses in a pipe subjected to both radial pressure and bending is illustrated in Fig. 1. For gravity flow pipe the net radial pressure is external which causes hoop compression stress (C) in the pipe wall. The pipe bending at section A-A adds compression stress (C) on the inner part of the pipe wall and tension stress (T) on the outer part. When the hoop compression stress exceeds the maximum bending stress, then the cross-section is entirely in compression. However, when the hoop compression stress is less than the maximum bending stress a portion of the pipe wall may have tensile stresses. When the net radial pressure is internal, the hoop stress is tensile. Bending will then increase the maximum tensile stress. This is the case for pipes for use with internal pressure (pressure pipe).

Pressure pipe design limits can be evaluated readily by applying internal pressure and holding it as long as desired. External end restraints can be used to control the longitudinal component of pressure. However the corresponding test for externally pressurized pipe is much more difficult. There are least two problems: 1) end restraints, and 2) buckling instability. General wall buckling could control the maximum hoop stress if external fluid pressure were used directly against the pipe. This would be unrealistic since the buckling strength of buried pipe is much greater when surrounded by soil than when surrounded by a fluid [2]. Local pipe wall buckling is also likely to be dependent on the characteristics of the surrounding material. The new test described in this paper overcomes these problems by applying load to the pipe through a surrounding ring of soil which supports the corrugation just as in an actual field loading.

The nature of radial pressure loading in the hoop compression tests is demonstrated in Fig. 2. This test produces only compression stress and strain in the pipe wall in the absence of effects such as local buckling and stress concentrations.

Principles of equilibrium show that the circumferential (hoop) thrust in the pipe wall is given by

$$T = PR \quad (1)$$

where

T = pipe hoop thrust per unit length of pipe (N/mm),

P = applied radial pressure (N/mm^2),

R = mean pipe radius (mm).

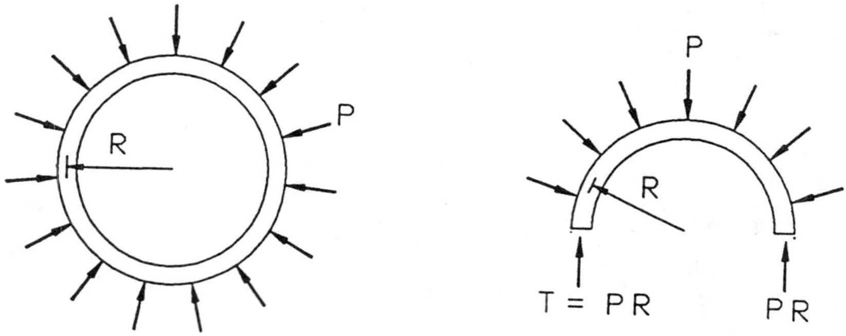

Fig. 2 Hoop Compression Test

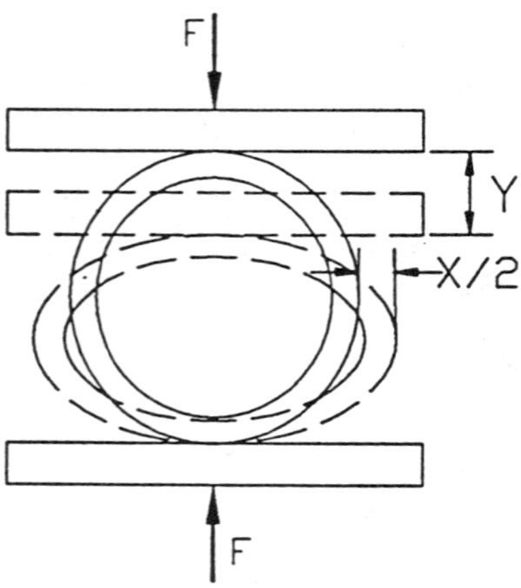

Fig. 3 Ring Bending Test

Hoop stress (σ, MPa) developed in the pipe wall by the radial pressure is

$$\sigma = T/A \ , \tag{2}$$

where

A = area of pipe wall per unit length of pipe (mm^2/mm).

The pipe modulus is defined as

$$E = \frac{\sigma}{[\Delta D/D_m]} \ , \tag{3}$$

where

E = modulus of elasticity of the pipe material (MPa),

D_m = mean pipe diameter (mm),

ΔD = change in mean pipe diameter (mm).

Equation 3 is valid because the average circumferential strain, $\Delta C/C$, is equal to the ratio, $\Delta D/D_m$. Substituting Equations 1 and 2 into Eq. 3 and rearranging gives as the expression for the change in pipe diameter

$$\Delta D = \frac{2PR^2}{E\,A} \ . \tag{4}$$

In analysis of plastics, the modulus of elasticity, E, in the above equations is time dependent when a specimen is under load and so the term viscoelastic modulus can be used. Details of viscoelasticity are discussed in detail in the ASCE Structural Plastics Design Manual [3]. The two typical conditions considered in viscoelastic materials are constant stress (creep) or constant strain (relaxation). In creep, the strains increase with time under load, and in relaxation the stresses decrease with time under load. In either case, Eq. 3 demonstrates that the viscoelastic modulus decreases with time under load. This continues indefinitely but at a rate that decreases on a logarithmic basis [3]. Creep and relaxation also occur in traditional construction materials such as wood and concrete.

All plastic pipe standards have a flexural pipe stiffness and strength requirement based on the ring bending test which is demonstrated schematically in Fig. 3; however, this test is not suitable for studying the pipe wall compression limits because it cannot produce hoop compression stress of sufficient magnitude to develop a limit state condition in the pipe. Although hoop compression and ring bending stresses are both present in actual pipe, the relative magnitude of each is highly variable depending on height of fill, soil stiffness, and uniformity of soil support. Thus it is not practical to have a combined

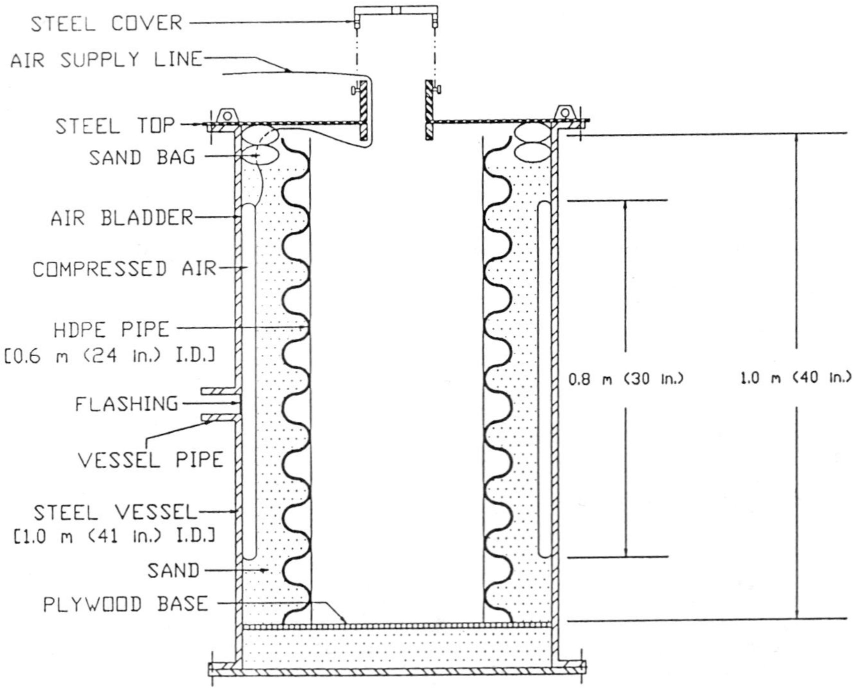

Fig. 4 Cross-Section of Pipe Hoop Compression Test Facility

test that addresses both stress modes. It is common in engineering to have separate tests for separate stress modes. The test discussed below is proposed as suitable for evaluating hoop compression to complement the ring bending test.

APPARATUS

The hoop compression test facility consists of a 900 mm (35.4 in.) diameter cylindrical steel pressure vessel, lined with an inflatable bladder (Fig. 4). The test section of pipe is installed at the center of the vessel, and the annulus between the pipe and the bladder is filled with tamped sand. Confined air in the bladder exerts radial pressure on the ring of sand around the pipe, resulting in circumferential compression of the pipe wall.

The air supply system uses a regulator to maintain constant pressure in the bladder for as long as desired. A pressure gage is used to monitor the pressure.

The bottom of the pressure vessel contains a layer of sand covered with a plywood sheet. The thickness of the sand can be adjusted to fit the length of pipe being tested.

The sand and air bladder are constrained by a steel top to resist the longitudinal (vertical) component of the air pressure. A hole is located in the steel top to provide access to the inside of the pipe during the test. This hole also has a cover.

Strains in the pipe wall can be measured by use of bonded electrical resistance strain gages. Diameter change is monitored by attaching extensometers (not shown in Fig. 4) between opposite sides of the pipe wall at various positions along the length of the pipe.

GEOMETRY OF TEST PIPE

The pipe used in the example test described in this paper was made of high density polyethylene (HDPE), cell classification 324420C as defined in ASTM Specification for Polyethylene Plastics Pipe and Fittings Material (D3350). It had annular corrugations on the exterior with a smooth interior liner heat bonded to the corrugations. The nominal interior diameter was 610 mm (24 in.). The exterior diameter was 726 mm (28.6 in.). The pipe was cut on cross-sections passing through the outside crest of the corrugations, to provide a length of 1020 mm (40 in.). The centroidal radius was 325 mm (12.8 in.) including the unbonded portion of the liner, and was 331 mm (13.0 in.) assuming that the unbonded portion of the liner was ineffective in supporting load. The profile cross-sectional area was 8.3 mm^2/mm (0.32 $in.^2/in.$) including the unbonded portion of the liner, and was 6.3 mm^2/mm (0.25 $in.^2/in.$) assuming that the unbonded portion of the liner was ineffective. Other tests, not reported in this paper, suggest that the liner is actually ineffective in resisting hoop compression loads because it is relatively thin and is only loaded at the edges where it is bonded to the corrugation. Fig. 5 shows the corrugation geometry and Table 1 shows average corrugation measurements. Note, that due to the manufacturing process, the corrugation wall thickness measurements (Nos. 3, 4 and 5) on one side of the corrugation were about 20% thicker than the measurements on the other side (Nos. 8, 9 and 10).

PROCEDURE FOR PIPE INSTALLATION

The sand layer at the bottom of the tank was adjusted so the distance from the plywood sheet to the bottom of the steel top was equal to or slightly greater than the length of the test pipe.

The bladder was centered vertically and taped against the inner vessel wall. Masking tape was used so that the bladder would remain held in place during the installation procedure, but would not be damaged upon bladder expansion. The pipe section was then placed in an upright position and centered in the vessel. The top and bottom of the bladder ended 130 mm (5 in.) from the pipe ends (Fig. 4).

The pipe was clamped in place during soil compaction by a bar placed across the top of the pipe. The bar was secured with a threaded rod passed through the pipe and anchored to the plywood base.

A moist, coarse to fine sand (SP) was used to fill the area between the bladder and the outer wall of the pipe. This region was about 83 mm (3.25 in.) wide. The sand was deposited between the outer wall of the pipe and the bladder in 200 mm (6 in.) thick layers. Each layer was compacted by hand tamping the sand using a 25 mm (1 in.) square by 1.2 m (4 ft) long piece of wood. This method achieved approximately 90% of maximum standard Proctor density (ASTM D698). The

Table 1 Average Corrugation Measurements

Thickness Measurement No.	(mm)	(in.)
1	7.09	0.279
2	3.89	0.153
3	4.27	0.168
4	4.42	0.174
5	4.01	0.158
6	2.87	0.113
7	2.95	0.116
8	3.38	0.133
9	3.53	0.139
10	3.53	0.139
11	3.91	0.154
12	2.72	0.107
13	2.77	0.109
Length Measurement Letter	(mm)	(in.)
A	76.96	3.030
B	24.05	0.947
C	47.88	1.885
D	45.75	1.801

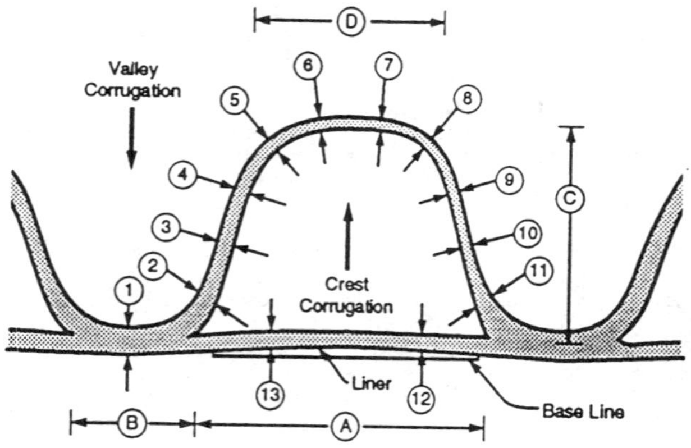

Figure 5 Corrugation Geometry

compacted sand completely filled the valleys in the outer pipe wall corrugations.

When the level of compacted sand reached the bladder air inlet, at the top edge of the bladder, the air line was attached. Then placing and compacting sand continued to a level about 76 mm (3 in.) from the top of the pressure vessel. The remaining space was filled with sand bags.

The steel bar and threaded rod used to temporarily secure the test pipe during the sand installation were disassembled and removed. Lastly, the steel top was bolted to the top flange of the vessel.

PROCEDURE FOR TESTING PIPE

Typically the air pressure in the bladder was increased in 34 kPa (5 psi) increments. Increments were applied quickly and then held for about 1,000 to 10,000 minutes (1 day to 1 week) to allow creep to develop.

The extensometers were read at the start of each increment and periodically during the creep process at constant bladder pressure. Because the rate of diameter change decreased with increasing time, the time interval between readings was increased from seconds at the beginning of the increment to a day at the end of the increment. Visual inspection was also made after each increment.

It was intended that the pressure would be increased until either the bladder failed or the pipe failed. In each of the tests to date, bladder leaks terminated the test. A pressure of 380 kPa (55 psi) has been achieved to date. According to measurements made in a deep burial field installation of a similar HDPE pipe, this pressure level represents more than 30 m (100 ft) of earth cover [4].

SAMPLE RESULTS

The changes in diameter with time for 7 successive pressure increments are shown in Fig. 6. The 4 sets of curves represent four extensometers arranged at 45 degree intervals around the pipe in the center 305 mm (1 ft) of pipe length. The uniformity of the 4 curves indicate that a uniform radial loading was achieved. The change in diameter at the end of each increment is shown as a function of the corresponding pressure in Fig. 7. The four extensometers are consistent with each other. No impending pipe failure is indicated at the highest pressure reached in this example, which was 241 kPa (35 psi).

The displacement versus time curves in Fig. 6 are replotted for each increment individually in Fig. 8 using log-log scales. Each increment is reasonably linear which is characteristic of viscoelastic materials. The slopes of the curves in Fig. 8 gradually increase with increasing pressure.

The data indicate 3% circumferential shortening and associated diameter decrease at the 241 kPa bladder pressure achieved in the test. The consequence in field installations is large positive arching or earth load shedding. Fig. 9 shows the displacement versus time for one extensometer and compares it with the predicted displacement assuming that all of the bladder pressure was transferred into the pipe. This was accomplished using the Boltzman superposition principle [3] and the

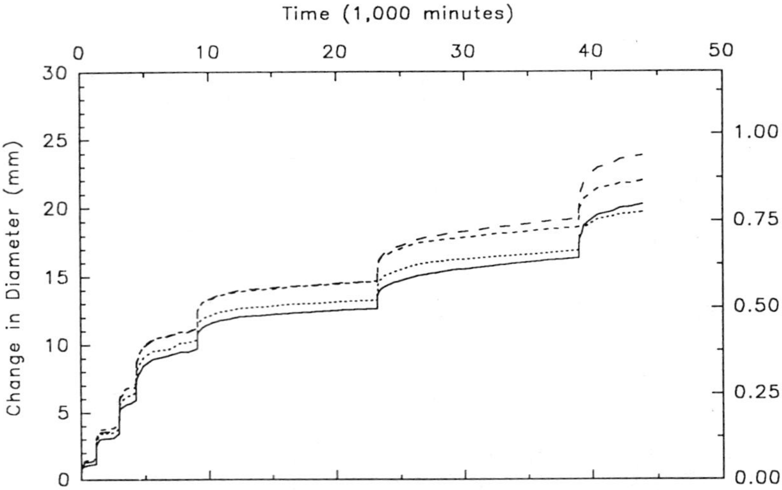

Fig. 6 Change in Diameter with Time

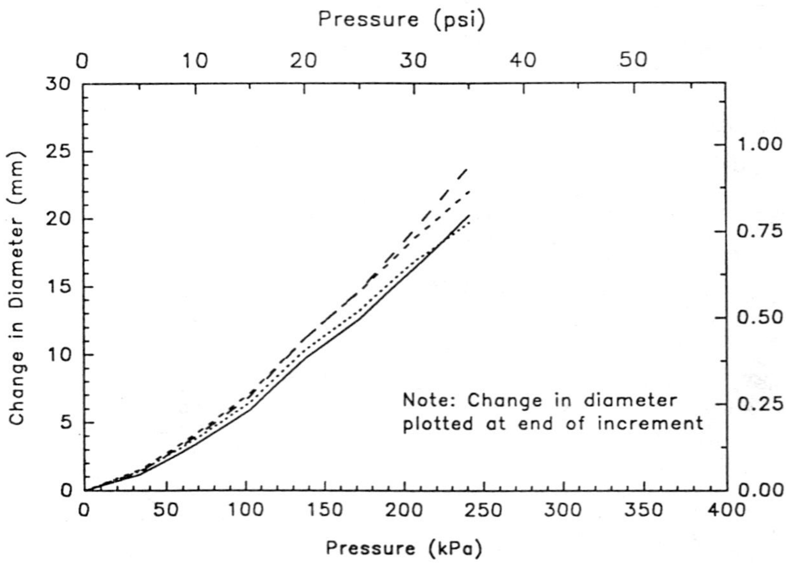

Fig. 7 Change in Diameter with Pressure

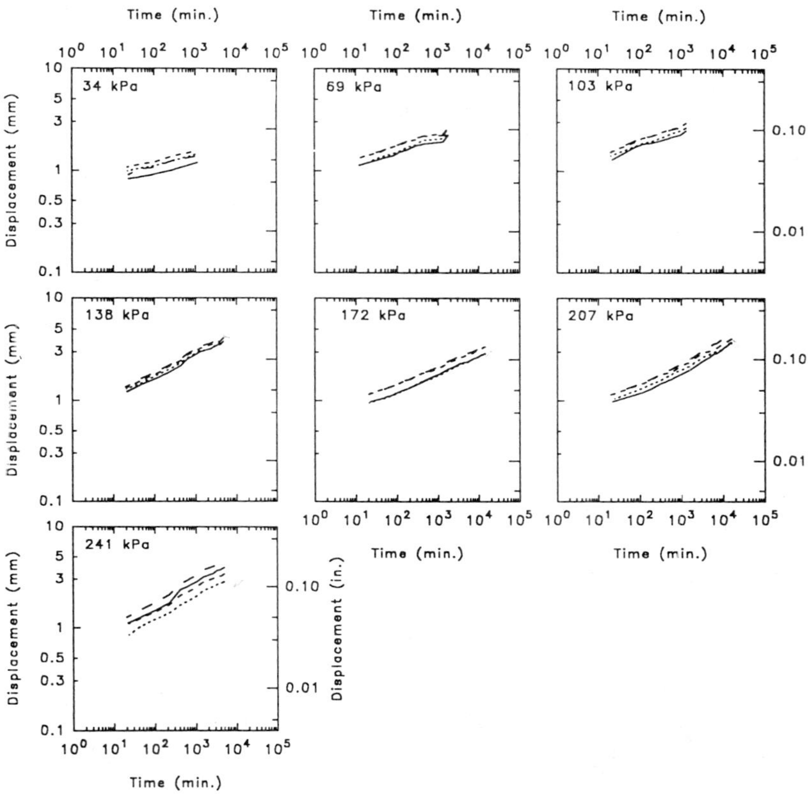

Fig. 8 Individual Displacement-Time Increments Plotted on Log-Log Scales

following modulus versus time curve developed by Hashash [4] for a similar HDPE corrugated pipe in ring bending:

$$E = 664\ t^{-.0859} \tag{5}$$

Where t is in seconds and E is in MPa. The difference between the two curves in Fig. 9 represents the portion of the load carried by the soil ring around the pipe. This demonstrates the beneficial arching that takes place as a result of the significant diametral shortening in a pipe with low axial stiffness, such as corrugated HDPE pipe. The actual in-ground load sharing that takes place is likely even more significant than demonstrated in this test because the extent of the soil is

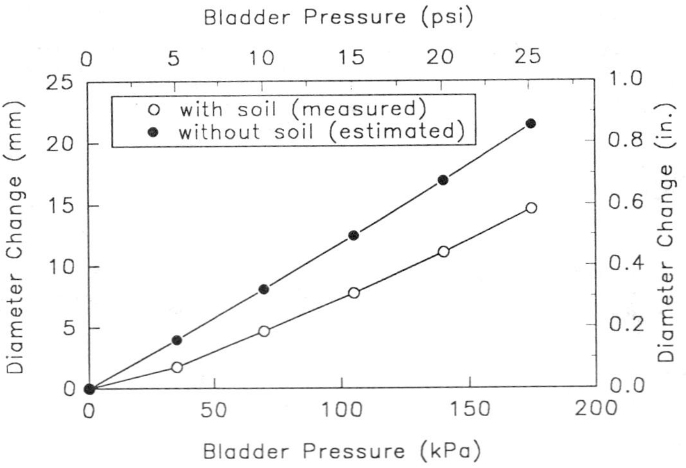

Fig. 9 Estimate of Load Carried by Soil

Fig. 10 Rippling of Unsupported Liner

limited in the test. This positive arching was demonstrated in a recent deep burial field test [4].

Rippling of the unsupported part of the liner was observed in the tests (Fig. 10) similar to that observed in the deep burial field test [4]. About 50 ripples occured around the full circumference of each corrugation. This was a result of the shortening of the corrugation to which the liner was attached. This phenomenon was first observed in the hoop compression tests when the radial pressure was about 138 kPa (20 psi) and the diametral shortening was about 1.9%. Since, as noted, the unbonded portion of the liner is largely inneffective in resisting stress, the rippling should not be considered a structural failure in and of itself; however, it can lead to high local stresses that may be significant in some designs. In the deep burial field test [4] rippling has been stable for over 6 years.

General pipe wall buckling did not occur during the tests even at 380 kPa (55 psi) bladder pressure. However, local buckling of the corrugated wall was observed after removal of the pipe which was subjected to 380 kPa (55 psi) pressure at the end of the test (Fig. 11). Also, the thinner corrugation wall (see Fig. 5 and Table 1) buckled while the thicker wall did not. The development of corrugation buckling could not be seen during the test because the pipe liner covered the corrugation. Local buckling was not observed in the tests with a maximum pressure of 241 kPa (35 psi). This phenomenon is a candidate for a wall compression limit. The ability to identify such effects, which are difficult to predict with mathematical models, demonstrates the benefit of the hoop compression test. Because the pipe specimen extends beyond the bladder (Fig. 4) the effects of the cut corrugation at the ends should be minimal, and thus the test should be equally applicable to annular profile wall pipe, such as reported here, and helical pipe. Observations of pipe deformation indicate that the central (400 mm) 16 in. of the pipe deformed uniformly, which also suggests that end effects were minimal.

CONCLUSION

A test has been developed which can subject a plastic pipe wall to high hoop compressive stresses representative of those in deep burial applications. This provides the ability to accurately determine the true wall compression limits relating to the effects of the material, the method of manufacturing and the profile shape. The test results should be directly applicable to design. The successful implementation of the hoop compression test of pipe specimens represents a significant new advance in pipe testing.

ACKNOWLEDGEMENTS

The hoop compression test was conceived by E.T. Selig and successfully implemented by L.C. DiFrancesco as independent research at the University of Massachusetts. The test was first used to help interpret results of a deep burial study of HDPE pipe sponsored by Advanced Drainage Systems, Inc. and the Pennsylvania Department of Transportation together with the U.S. DOT Federal Highway Administration. Further experience with the test was achieved as part of a study of compression limits of HDPE pipe conducted for the Corrugated Plastic Pipe Association. Valuable assistance was provided by Dr. J.A. Donovan, Associate Professor of Mechanical Engineering, by co-author T.J. McGrath, and by Walter Clark, Charles Cichanowicz and Miles Eastman who are Univ. of Mass. College of Engineering technicians.

REFERENCES

[1] Standard Specification for Design of Highway Bridges, American Association of State Highway and Transportation Officials, 15th Edition, Washington, D.C., 1992.

[2] Moore, I.D. and Selig, E.T., "Use of Continuum Buckling Theory for Evaluation of Buried Plastic Pipe Stability," Buried Plastic Pipe Technology, ASTM STP1093, Buczala and Cassady, Ed., October 1990, pp. 344-359.

[3] Heger, F.J., Chambers, R.E., and Dietz, A.G., "Structural Plastics Design Manual." ASCE Manual of Practice No. 63, ASCE, New York, 1984.

[4] Hashash, N. and Selig, E.T., "Analysis of the Performance of a Buried High Density Polyethylene Pipe," Proceedings; of the First National Conference on Flexible Pipes, Columbus, Ohio, October 1990, pp. 95-103.

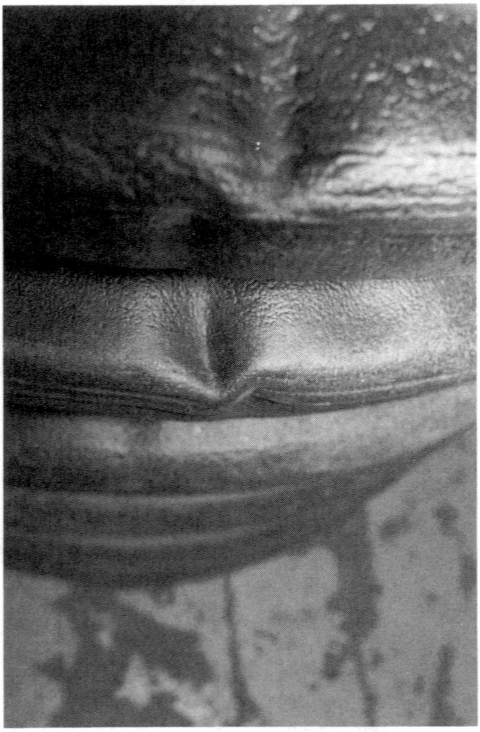

Fig. 11 Local Buckling of Corrugation Wall

Patrick S. Leevers[1], Grigorios P. Venizelos[2], and Robert E. Morgan[2]

RAPID CRACK PROPAGATION ALONG PRESSURISED PLASTIC PIPE

REFERENCE: Leevers, P. S., Venizelos, G. P., and Morgan, R. E., "**Rapid Crack Propagation along Pressurised Plastic Pipe**", Buried Plastic Pipe Technology: 2nd Volume, ASTM STP 1222, Dave Eckstein, Ed., American Society for Testing and Materials, Philadelphia, 1994.

ABSTRACT: Rapid Crack Propagation (RCP) is a rare but dangerous failure mode in gas and water pipelines. Although relatively low RCP resistance may be endemic to plastics, extruded polyethylene pipe varies widely in the *critical pressure* below which the arrest of RCP is ensured at each temperature. The proposed International Standard 'S4' pipe test for RCP, and the prediction of S4 and full-scale test critical pressures from basic material data, are discussed. A more effective quality index may the temperature above which and the wall thickness below which RCP becomes impossible at *any* sustainable pressure. This brittle-tough transition in polyethylene also appears in dynamic fracture resistance tests as a minimum stable crack velocity. Although the well known Irwin-Corten expression can be used to predict critical pressures for notched, water pressurised pipe, the proximity of this transition velocity to that at which the crack driving force in gas-pressurised pipe is a maximum may preclude a similar approach. Charpy-type tests, on sharp-notched specimens of the base resin at varying temperatures and thicknesses, can be used to locate the transition sufficiently closely to index resins. Predicting the RCP performance of pipe extruded from these resins, however, remains uncertain.

KEYWORDS: polyethylene, pipe, RCP, test, dynamic, fracture, toughness, transition, thickness, temperature, Charpy

INTRODUCTION

Full-scale tests on buried pipelines indicate that above a critical pressure p_c, Rapid Crack Propagation (RCP) along a polyethylene pipeline pressurised by gas or an air-water mixture can be sustained indefinitely [1]. As a service failure mode, RCP is undoubtedly rare, but failures have occurred (although information on them is hard to come by) in fuel gas pipelines under pneumatic proof testing. Failures have also occurred in pressurised water distribution pipelines in service, but these, too, have been rare despite the higher service pressures generally used. Full scale tests confirm that the critical pressure for water pressurisation is much higher than for gas. Tests on 100% water filled MDPE pipe could not sustain RCP at any attainable pressure, but a

[1] Senior Lecturer, Mechanical Engineering Department, Imperial College of Science, Technology and Medicine, London SW7 2BX, United Kingdom

[2] Graduate Student, Mechanical Engineering Department, Imperial College

10% entrapped air volume was sufficient almost to reduce p_c to the 100% air level [2].

The RCP threat is perceived differently from opposite sides of the Atlantic. In the USA a pragmatic approach has prevailed. Surveying distribution networks in which pipe diameters are generally small and pressures generally low, utilities have assessed the RCP risk, according to the incidence of failure, to be negligible. However, there is also a perception that slow crack growth – the main source of field failures in polyethylene (PE) pipe – has been beaten. This has brought the temptation to exploit PE in areas of the pressure/diameter envelope which had previously been avoided.

In Europe, utilities have generally retained an awareness that RCP may pose a potent threat in these unexplored areas even for otherwise reliable materials; its low probability is balanced against the enormous destruction threatened by a large-scale sustained RCP failure, not only in the physical domain (especially for a fuel gas pipeline) but also for the credibility of what many still regard as unconventional 'engineering' materials. It is therefore important to recognise that RCP is more improbable in some materials than in others. British Gas has for some time specified full-scale RCP testing as a condition of type approval, but with the introduction of new 'third generation' materials, promising sustained service up to still higher pressures and diameters, vigilance has redoubled: currently, *every batch* of this material is subjected to a full scale test.

This awareness has led to agreement to include an RCP test within forthcoming ISO (International Standards Organisation) and CEN (Comité Européen de Normalisation) standards for PE fuel gas distribution pipe: specifically, the 'S4' (Small Scale Steady State) test [3], measuring resistance to RCP in a pneumatically pressurised section of as-produced pipe. The S4 method, one of an evolutionary line originating with the wide-plate Robertson crack arrest test, aims to sustain the steady state observed in full-scale tests within a specimen only 7 diameters long. Since it does so by suppressing the axial gas decompression process which, in a long pipe, allows partial unloading before the crack front arrives, the S4 critical pressure is lower by a factor of at least 3.

The proposed use of this method for type or quality control testing will oblige the specifying engineer to consider the RCP critical pressure at 0°C as a quality index as important as the 50 year failure stress. For material developers, this brings a need to predict the RCP performance of pipe from material properties measured using small specimens, prior to extrusion of pipe in batch quantity. This paper reviews our research on two strategies for doing so: direct prediction of the critical pressure using fracture mechanics; and total avoidance of the RCP regime using a 'ductile-brittle transition' approach.

PREDICTION OF CRITICAL PRESSURE USING LEFM

Even in the highly ductile materials used for pipe, the brittle nature of RCP has always encouraged the application of Linear Elastic Fracture Mechanics (LEFM) to predict the critical pressure. This involves three distinct steps: measurement of the dynamic crack propagation resistance G_D of the material; calculation of the crack driving force G developed by the structure, as a function of pressure; and solution for the critical pressure to sustain propagation by setting

$G = G_D$. Irwin and Corten, who first applied this approach, considered the crack to be driven entirely by unloading of the circumferentially stressed pipe wall. This provides a driving force

$$G_0 = \frac{\pi p_0^2}{8E}(D-t)\left(\frac{D-2t}{t}\right)^2 \tag{1}$$

where p_0 is the initial pipeline pressure, D and t are the pipe diameter and thickness (the Standard Dimension Ratio, SDR, being their nominal ratio) and E is the tensile modulus of its material (which for polymers must be measured at an appropriate rate). Setting $G_0 = G_D$ led them to predict a critical pressure of

$$p_c = \frac{1}{\mathrm{SDR}-2}\sqrt{\frac{8EG_D}{\pi D}\left(1-\frac{1}{\mathrm{SDR}}\right)}. \tag{2}$$

For gas and water pressurised PE pipe, none of the steps in this procedure is straightforward.

The brittle RCP mode in PE is dynamically stable only at crack speeds greater than about 100 ms^{-1}, below which a higher resistance tearing mode predominates. Any G_D test method must therefore provide the power to drive a crack front through the high initiation resistance very quickly, and yet allow it to stabilise rapidly enough, and then for long enough, to observe during stable propagation – all within a small specimen. Another size restraint is imposed by the high toughness to yield-stress ratio of PE, which according to the ASTM Test Method for Plane-Strain Fracture Toughness of Metallic Materials (E 399) would require a specimen at least 40-50 mm thick to achieve plane strain conditions. Finally, any specimen in which large strains are induced must account for pronounced elastic non-linearity.

On the other hand, calculation of G for a pressurised elastic shell is notoriously difficult. Even in the absence of any 'external' work contribution from the pressurised fluid, a full three-dimensional elastodynamic deformation analysis is required. For a gas-pressurised pipeline, the problem threatens to become indeterminate, since the relatively small energy exchanges which determine crack behaviour take place in the presence of an effectively infinite energy source.

Dynamic crack resistance of PE and other tough polymers

Neither of the material properties E or G_D which appear in Eq 2 is easy to measure. So rate dependent is the modulus of PE that barely one third of the strain energy stored in the pipe wall can be recovered from unloading around a fast-running crack, whilst it is now widely accepted that Charpy tests, even if carried out on V-notched specimens and analysed on a energy-density basis, are a poor guide to the dynamic crack propagation resistance G_D.

However, the High Speed Double Torsion (HSDT) test developed by Leevers and Wheel [4, 5] allows G_D to be measured during crack propagation at almost constant speeds of 100-300 ms^{-1} and more. Computation of G_D from displacement and crack length readings involves elastodynamic analysis, which itself, for nonlinear materials, requires E *vs.* strain data. These are provided by an supplementary procedure involving simulated HSDT tests on dummy pre-fractured specimens [6]. In the small-strain limit, the results from this procedure agree well with those from 1 MHz ultrasonic wave propagation methods.

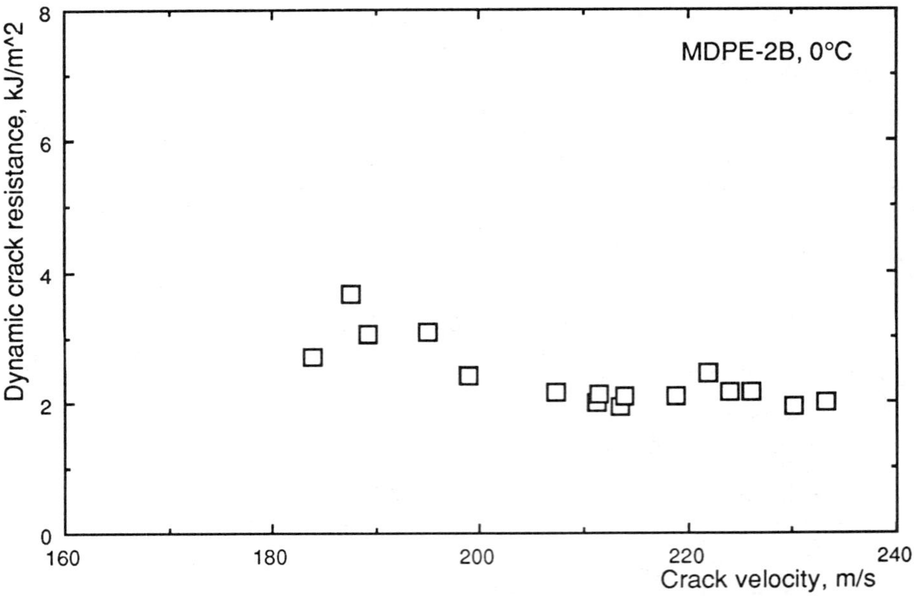

FIG. 1--Dynamic crack resistance G_D of MDPE-2B at typical pipe RCP velocities, measured using the High Speed Double Torsion test

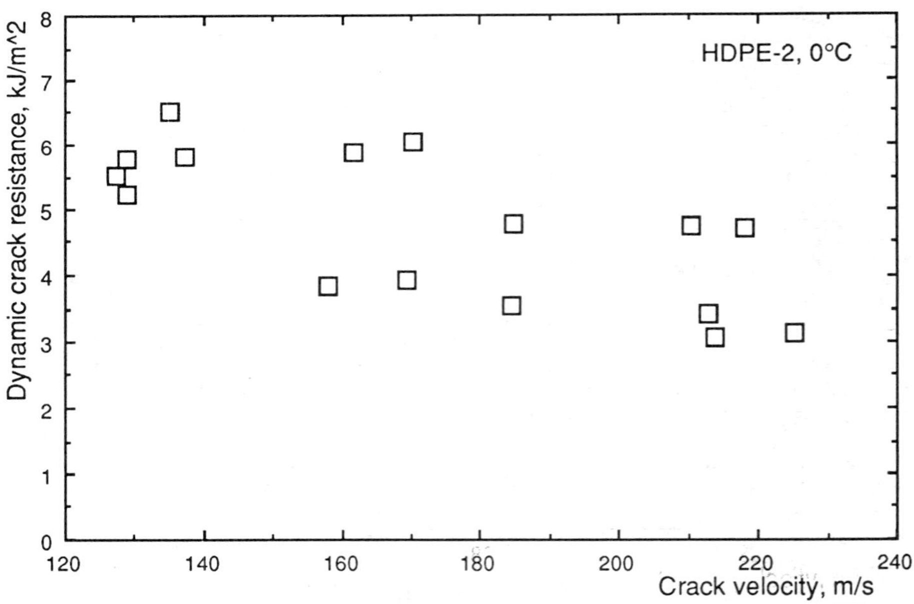

FIG. 2--Dynamic crack resistance G_D of HDPE-2 at typical pipe RCP velocities, measured using the High Speed Double Torsion test

HSDT results for two representative resins are shown in Figs. 1 and 2. MDPE-2B is a blue (water grade) commercial medium density PE whose S4 test performance ranks it near the top of its class. HDPE-2 is one of the new, high-performance 'third generation' materials, based on modified high density PE resins, whose RCP performance is distinctly better: in fact, no batch of material yet tested has sustained RCP in an S4 test at 0°C. The most surprising result from HSDT testing, therefore, is that RCP is sustained in HDPE-2 at all. We will show that this is almost certainly attributable to thickness effects. Because the HSDT crack front does not remain straight and orthogonal to the plate surface but is curved and propagates self-similarly at a shallow angle to it, its length is much greater than the plate thickness. For the geometry used here, 100 mm × 200 mm × 6-10 mm thick, the physical crack front length comfortably satisfies the ASTM E 399 plane strain criterion, and Wheel has demonstrated [6] that thickness effects are indeed absent.

Two further features of these results are particularly important. Firstly, increasing crack velocity reduces G_D towards a plateau whose value is surprisingly low: about 2 kJ m^{-2} for MDPE and 4 kJ m^{-2} for HDPE. Secondly, there is a crack speed below which dynamically stable propagation cannot be sustained; crack extension degenerates into the stick-slip mode classically associated with a downward-sloping G_D *vs.* crack speed characteristic.

Application of the Irwin-Corten model to water pressurised pipe

The assumptions on which the Irwin-Corten model is founded are undermined by the presence of the pressurising medium itself. However, for a liquid of high bulk modulus *M* such as water, the increase in strain energy content is modest: if all of it could be recovered Eq 1 would become

$$G_0 = \frac{\pi p_0^2}{8E} D(\mathrm{SDR}-2)^2\left[1+\left(\frac{E}{M}-1\right)\frac{1}{\mathrm{SDR}}\right], \tag{3}$$

and the critical pressure would be

$$p_c = \frac{1}{\mathrm{SDR}-2}\sqrt{\frac{8EG_D}{\pi D}\frac{\mathrm{SDR}}{\mathrm{SDR}-1+\dfrac{E}{M}}}, \tag{4}$$

which for water, whose bulk modulus is roughly equal to the dynamic tensile modulus of PE, and for practical SDR's exceeding 10, differs little from Eq 2.

Classically, LEFM strength prediction uses plane-strain data in order to identify the worst case. Here the HSDT test yields the required data, strictly appropriate only to very thick pipe walls; but pipe-grade PE grades seem to differ little in plane strain crack resistance, whilst they are so tough in plane stress that RCP is impossible and LEFM inapplicable. The observation of a 1-2 mm wide plane-stress tear zone at the bore on RCP fracture surfaces in HDPE-2 led us to test pipes 'sidenotched' by a full-length, axial, internal surface score of similar depth in order to suppress it. Air pressurised S4 tests on these specimens demonstrated a spectacular reduction in critical pressure [6]. Since the notch removes little strain energy (i.e. its stress intensity factor is insignificant compared to that of the axial crack) it is reasonable to identify these results as true plane-strain fractures.

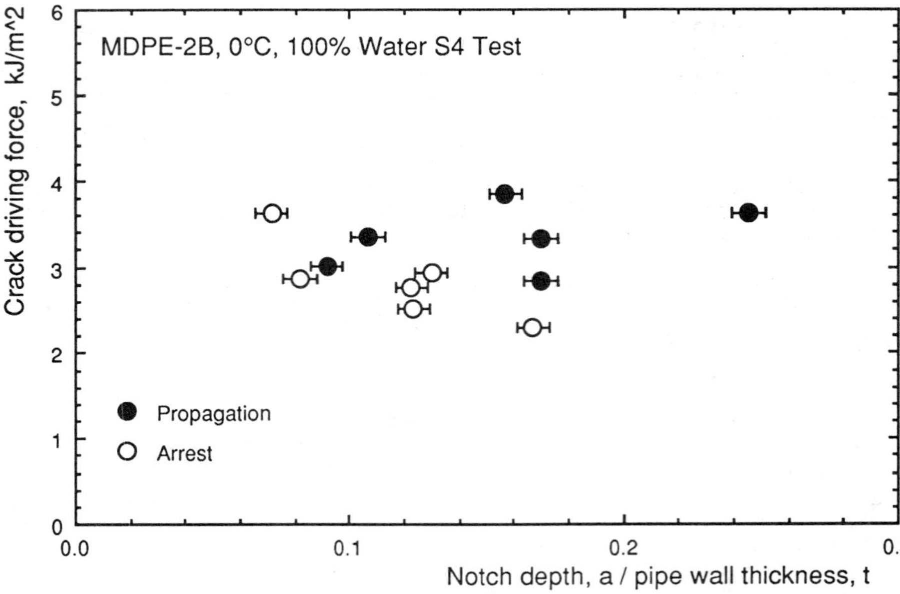

FIG. 3--Response of the crack at various driving forces and notch depths in water-pressurised S4 tests on sidenotched MDPE-2B pipe (125 mm SDR11)

It follows that the modified Irwin-Corten equation (Eq 4), using the plane strain HSDT value for G_D, should correctly predict the critical pressure of a sufficiently deeply sidenotched water-pressurised pipe. We have recently carried out tests on axially notched MDPE-2B 125 mm SDR 11 pipe pressurised by water at 0 °C. Fig. 3 shows the results, classified only as 'arrest' or 'propagation' (the proposed ISO Standard S4 method embodies a definition, but the transition is usually very clear cut anyway [3]). The crack driving force *G*, calculated from Eq 3 multiplied by a factor of $t/(t-a)$ to correct for the reduced crack path width, is quite consistent with the data of Fig. 1 for sidenotch depths *a* exceeding about 1 mm.

We believe these results to be the first direct confirmation of the Irwin-Corten theory for PE pipe. They are bracketed by special conditions: water pressurisation emphasises the importance of occluded gas to the driving force for RCP in pipe, whilst sidenotching emphasises the importance of the plane-stress to plane-strain transition to the RCP resistance of the material. We will discuss both of these issues, turning immediately to the first.

<u>Gas pressurised pipe</u>

Direct application of the Irwin Corten theory to gas pressurised pipe predicts a dangerously non-conservative p_c. This is because pressurised gas can contain a thousand times as much energy as water under the same pressure; it is mainly the dynamics of this vast energy source which determine G. There are three strong motives to compute G for both full-scale and S4 air-pressurised pipe tests:

1 to underpin the factor of 3 to 3.5 which relates S4 and full-scale critical pressures, by analysing and comparing G in both geometries;
2 to interpret S4 test results as a pipe material property (as shown above for water pressurised tests) and hence, ultimately
3 to model crack propagation and arrest from a knowledge of the material's dynamic fracture resistance G_D, and hence to the predict critical pressure for a full-scale pipeline.

As a crack runs along a pipeline, fluid escapes through it. Before a crack front propagating at a velocity $\dot{a}$ (less than the sonic velocity C_0) arrives at any fixed cross-section, therefore, a decompression wavefront sweeps past and the pressure begins to decay from p_0 to some proportion of it which depends only on $\dot{a}$. The pipe wall therefore unloads, dumping strain energy into the fluid. According to a simple one-dimensional analysis, the crack tip pressure will fall to

$$\frac{p_1}{p_0} = \left[1 - \frac{\gamma - 1}{\gamma + 1}\left(1 - \frac{\dot{a}}{C_0}\right)\right]^{2\gamma/(\gamma-1)} \tag{5}$$

and it can be argued that replacing p_0 in Eq 1 by p_1 would provide a better estimate of the reduced crack driving force.

However, the remaining pressure does not collapse immediately behind the crack tip, but decays steadily over a characteristic length. Within this *outflow* zone, the exhausting gas develops a high crack driving force by expanding against the pipe flaps. The crack driving force G is the total (strain plus kinetic) energy per unit pipe length lost between planes ahead of and behind the crack front, plus the total forward axial force exerted by pressure against the flared-out walls, per unit crack width (i.e. wall thickness). It is the need to accurately compute this pressure component, which depends on interactions of fluid dynamics of the gas and elastodynamics of the pipe wall, which rules out a closed analytical solution of the Irwin-Corten form.

The most advanced computational method yet adopted to calculate G, due to O'Donaghue and Kanninen [7, 8] resorts to dynamic finite-element simulation of transient axial crack growth from zero length. G, computed at each time step, rises steadily as the crack extends before stabilising at a plateau which is assumed to represent the steady state. For a fully coupled fluid-structure dynamic model, the computation becomes a large and expensive task, and must usually be simplified by

1 uncoupling the fluid analysis (after having computed, using the coupled formulation, a few representative pressure decay profiles which are then generalised); and
2 terminating the simulation after a few diameters of crack growth — at which (particularly for high crack speeds) it may not be clear that the plateau truly represents the steady state G.

This plateau value, plotted as a function of crack velocity, exhibits a peak at an intermediate speed. For slower cracks G is reduced by gas decompression and outflow, whilst for faster cracks it has been depleted by inertial effects which divert pressure work into kinetic energy.

We have developed an alternative approach based on *a priori* assumption of a steady state, and using the Finite Volume method rather than the more common Finite Element method. The formulation allows G to be computed for a moving control volume extending from a plane ahead of the crack tip, at which the pipe wall is in equilibrium with the partially decompressed fluid contents, to a plane behind the crack tip at which the fluid has completely escaped. Pressure is assumed to decay linearly from p_1 computed *via* Eq 5 to zero (atmospheric) pressure over a decompression length L. Whilst inertial effects are fully accounted for, this steady-state approach eliminates time from the problem, allowing the crack driving force G to be calculated, for a given pressure profile, from a 'snapshot' evaluation of the displacement field.

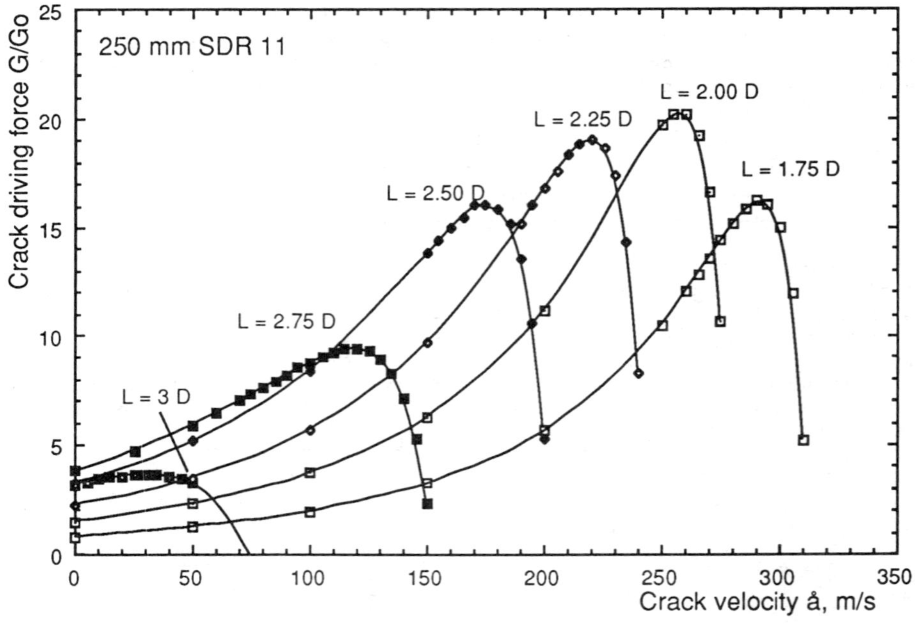

FIG. 4--Crack driving force G in a gas pressurised pipe for various decompression lengths, L

Fig. 4 shows G (normalised against the Irwin-Corten value) *vs.* $\dot{a}$ results, computed by Finite Volume analysis of a $6D$ long control volume of 250 mm MDPE SDR 11 pipe for various decompression lengths L. Although G depends strongly on L, attaining a maximum at a different crack speed for each value studied, the envelope traced by these maxima probably provides a close approximation to the maximum G for *any* realistic decompression profile. This maximum G is, as expected, much greater than G_0.

Now, comparing Figs. 4 and 1 exposes a problem which arises at the third step of LEFM analysis, when G and G_D are equated — which in this situation, with both being functions of crack velocity $\dot{a}$, involves overplotting Fig. 4 at different pressures onto Fig. 1 and seeking intersections. G scales linearly with pressure, and if the G peak were to touch a flat $G_D(\dot{a})$ characteristic as assumed by Kanninen *et al* [8] this process would be well-posed. In fact, the velocity at which G for gas-pressurised pipe attains an absolute maximum is not far from that below which steady RCP cannot be sustained. The assumption that the decompression length L adjusts itself to maximise G becomes weaker as $\dot{a}$, and the inertial constraints on gas outflow, increase. With G falling more precipitously from a peak at a lower speed, as the results computed by Kanninen *et al.* [8] suggest, the uncertainties of locating the $G = G_D$ intersection may become overwhelming. In some ways the problem is worse for the S4 test, in which G has a maximum near zero and falls monotonically and even more steeply with increasing crack velocity.

BRITTLE TOUGH TRANSITIONS IN PIPE-GRADE PE

Thus the minimum crack velocity below which increasing resistance prevents steady RCP in dynamic fracture resistance tests on PE, manifesting a 'brittle-tough' transition, may be at least as important as the lower-shelf G_D value. Further research using the S4 test also shows that a more discriminating index of performance may be the temperature above which (and the wall thickness below which) RCP becomes impossible at *any* sustainable pressure.

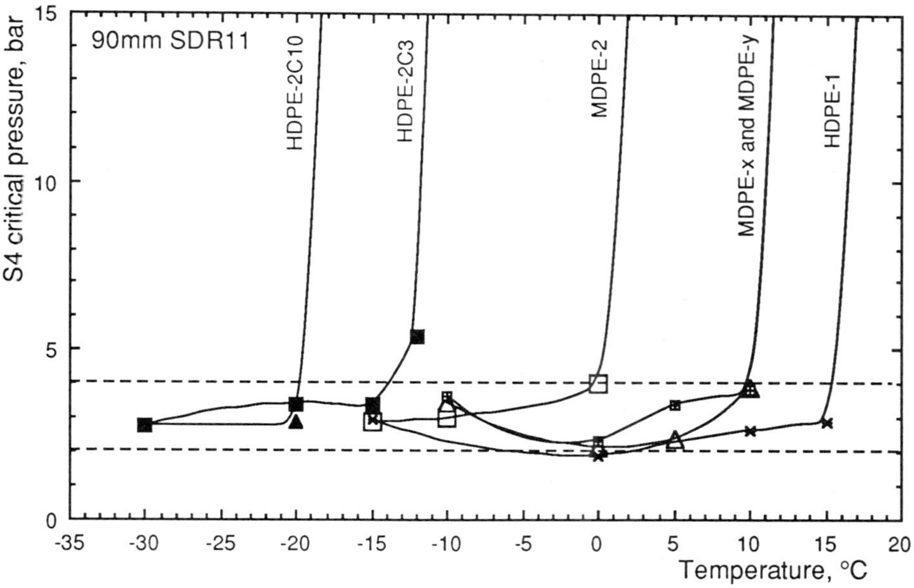

FIG. 5--Ductile-brittle transition temperatures for 90 mm SDR11 pipe of various PE grades

Temperature effects in the S4 test

S4 test critical pressures vary strongly with temperature, falling at low temperatures to a plateau which differs little between very diverse materials. The critical pressure sweeps upwards quite abruptly at a temperature which can be characterised (for several pipe sizes) by that at which RCP cannot be sustained at 10 bar. The fracture surfaces show that this temperature corresponds to a ductile-brittle transition (DBT).

Fig. 5 shows a critical pressure *vs.* temperature plot for several very different resins (including 'first generation' HDPE-1 as well as MDPE-2 and HDPE-2 grades) tested as 90 mm SDR 11 pipe. Larger diameters show a similar trend, but, for a given resin, the lower-shelf critical pressure is lower, the DBT temperature higher, and the transition gentler.

The narrow range of lower-shelf critical pressures recalls the relatively minor difference between plane-strain G_D data for MDPE-2 and HDPE-2 (and other grades of PE and even other semi-crystalline thermoplastics) at 0°C; the principal difference between resins lies in their DBTT, which varies from −17°C for the best HDPE-2 to room temperature for HDPE-1. This parameter provides not merely a pass/fail criterion but a milestone from service conditions to the RCP danger zone. For buried water pipelines, RCP tests should seek to establish whether the transition temperature is greater or less than 0°C. For fuel gas applications the situation is less clear cut: although temperatures are generally a little higher, cold spots can develop due to adiabatic decompression at diametral expansions.

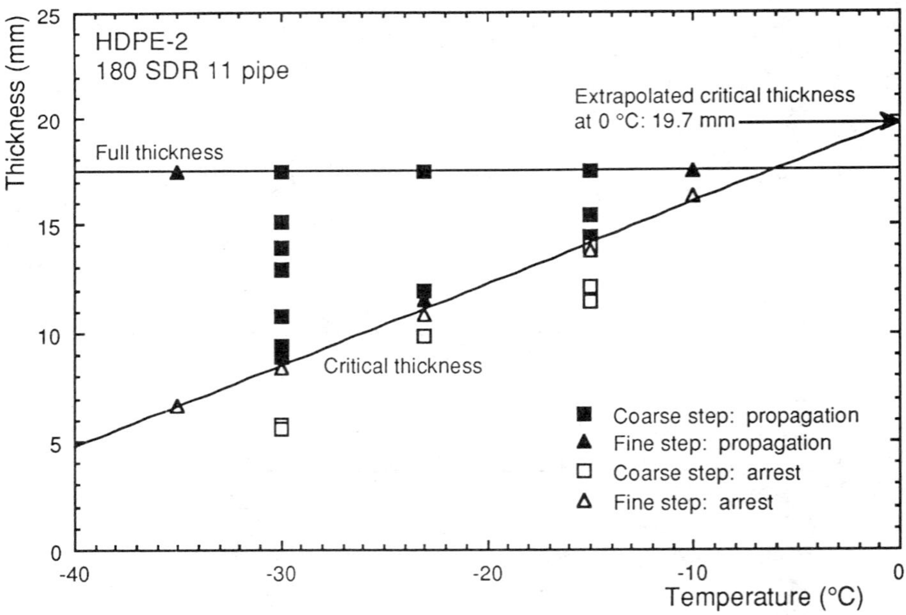

FIG. 6--Variation of critical thickness with temperature for 180 mm SDR11 HDPE-2 pipe

Thickness effects in the S4 test

Another modified S4 test procedure [6] demonstrates a *critical thickness* below which RCP cannot be sustained at any pressure. The outer diameter of each pipe specimen is machined to reduce the wall thickness in a series of one-diameter long steps. On RCP testing at 50% above full-thickness critical pressure the crack arrests at a characteristic thickness, t_c, which is evaluated by refining the step height from test to test. This procedure has been applied successfully to 90, 180 and 250 mm SDR11 pipes.

Using this procedure, we have investigated the temperature dependence of t_c. Original results for HDPE-2 180 mm pipe in which RCP was unattainable at 0°C (Fig. 6) indicated that t_c depended linearly on temperature, encouraging extrapolation to t_c = 19.7 mm at 0 °C and DBTT = -6 °C at 17.5 mm (full thickness). However, S4 tests on 250 mm SDR 11 pipe (t = 23.5 mm) and full scale tests on 500 mm SDR 11 pipe (t = 45 mm) could not sustain RCP at 0°C, suggesting that this extrapolation was over-conservative. Further evidence reveals that the t_c locus is linear only over a certain temperature range and tends to climb more abruptly as the full thickness DBTT of the pipe is approached.

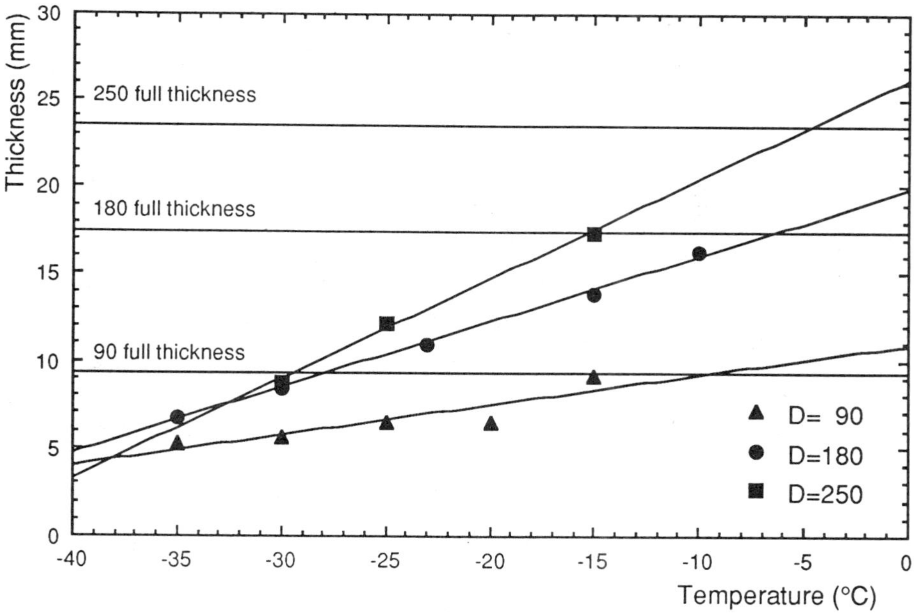

FIG. 7--Critical thickness loci for three different HDPE-2 pipe diameters

Since the danger of straying into the locus of RCP susceptibility at 0°C by using larger diameter HDPE-2 pipes at higher pressures clearly cannot be discounted, it was explored through critical thickness tests on 90, 180, and 250 mm SDR 11 pipes. Fig. 7 presents the t_c arrest loci from these tests, showing that the extrapolated t_c at 0 °C increases with pipe diameter in such a way that it always lies above the full thickness of the pipe. Clearly, t_c is not purely a material property; it also depends on the effect of diameter on the crack driving force developed in a pipe of a given thickness (providing a further motive for computing *G*). Although this dependence keeps the danger of RCP at 0°C one step further away for each increase in pipe size, the estimated full thickness DBTT seems to edge towards 0°C. This trend is in agreement with the fact that, for a given resin, the temperature at which the S4 critical pressure for RCP sweeps abruptly upwards (as in Fig. 5) is higher for larger diameters, and reaffirms the need for caution in using larger diameter or thicker HDPE-2 pipe at higher service pressures.

Two other issues have been explored using these methods. Firstly, as the pipe is machined down, the residual stress profile changes. The pipe wall is stress relieved as material closer to the outer surface, and under high circumferential and axial compression, is removed. Tests on annealed and non-annealed pipes have demonstrated t_c loci that converge at low temperatures where t_c values are low, and diverge at higher temperatures as t_c approaches the full thickness, for which the residual stresses are highest in non-annealed pipe (Fig. 8).

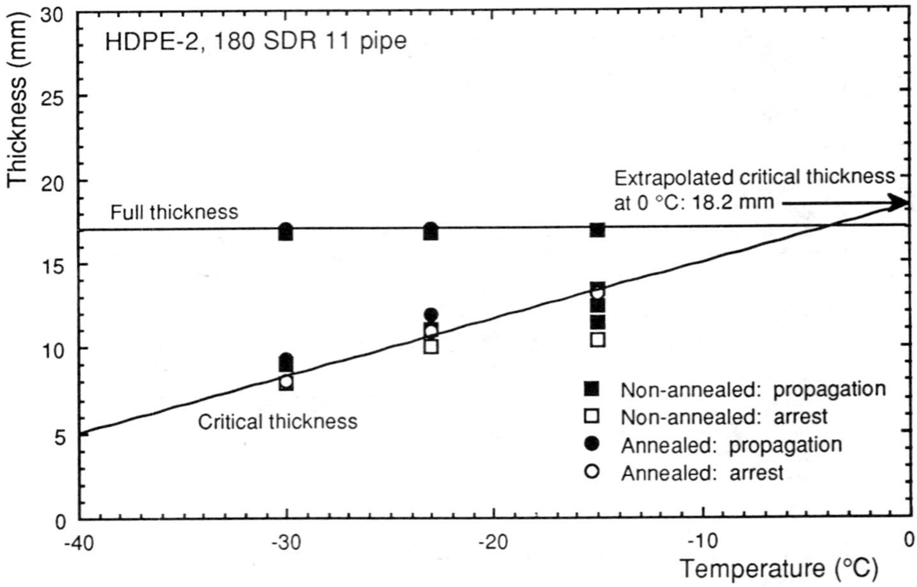

FIG. 8--Effect of residual stress on the critical thickness t_c of HDPE-2 180 mm SDR 11 pipe

Secondly, there is some evidence that critical thickness is independent of the variation in crystallinity through the pipe wall thickness. Tests on pipe machined from the inside (removing material of the highest crystallinity) and from the outside (removing material of the lowest) have indicated similar t_c values to within the machining tolerance.

Use of Charpy tests to locate the brittle-tough transition

For the best part of a century, the method of first recourse for locating a DBT has been the Charpy test. This remains the most accessible quality control method for most PE pipe users, providing a strong motive to develop a Charpy-based procedure for locating the critical thickness/diameter locus.

The effect of thickness on DBTT was measured using sharp-notched instrumented Charpy tests on specimens 55 mm long, 10 mm deep and varying in thickness B from 4 to 16 mm. The span was 40 mm and the impact speed 1.5 ms^{-1}, giving a strain rate of 56 s^{-1}, in all tests. Specimens were tested across a range of temperatures from −70 to +23°C. Conventional instrumented impact test readings provide both peak load and total energy to failure. Calculation of an apparent toughness both 'legally' (from peak load) and 'illegally' (from total energy) identifies a clear nil-ductility temperature above which they diverge (Fig. 9). This accurately reflects the fracture surface DBT temperature, making subjective 'percentage ductility' evaluations unnecessary. Note how much greater the lower bound for G_C is than G_D plotted in Fig. 2; we have as yet no way to distinguish thickness effects from those of initiation transients.

Plotting the DBTT *vs.* inverse thickness reveals a linear relationship (Fig. 10). The superior performance of HDPE-2 over MDPE-2 is indicated by a lower intercept on the temperature axis, representing a lower 'infinite thickness' DBTT. However, these results suggest that the DBTT of HDPE-2 is much more sensitive to thickness than that of MDPE-2, once again emphasising the care needed in using large diameter HDPE-2 pipes at high pressures, since a thickness promoted ductile-brittle transition may lift the DBTT above the minimum service temperature. However, these results should be interpreted with care, since the DBTT is also known to be very sensitive to strain rate. The extrapolated 'infinite thickness' DBTT is therefore only relevant to a particular Charpy test geometry and strain rate, but may still be useful for comparing different resins.

Fig. 11, for HDPE-2, also shows temperature *vs.* inverse critical thickness from S4 pipe tests – i.e. data of the type shown in Fig. 6. The lines have similar slopes, although the temperature intercepts differ (probably due to strain rate effects). This suggests that it may be possible to use the Charpy test to produce a master curve for each resin, from which a DBTT for each pipe size could be predicted using a shift rule.

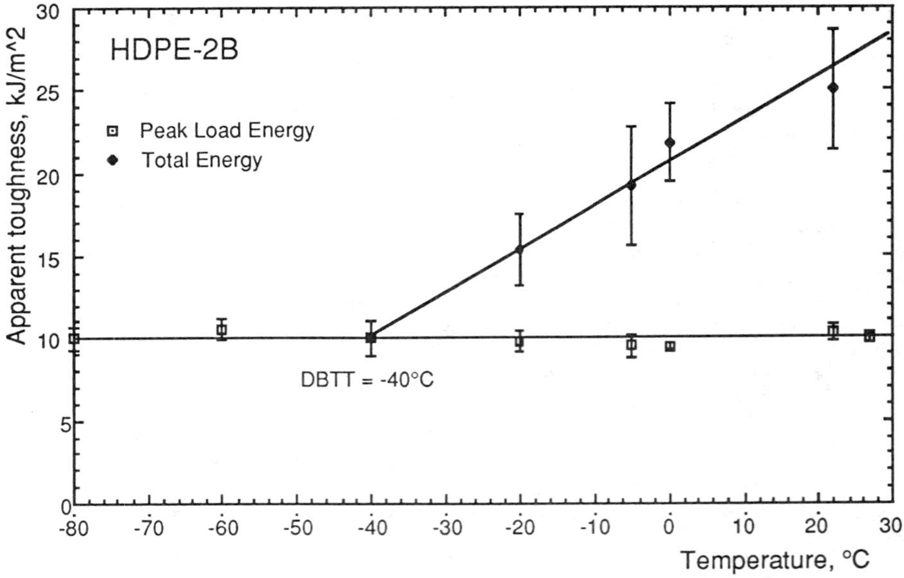

FIG. 9--Definition of the DBTT for an HDPE-2 grade from modified Charpy test data

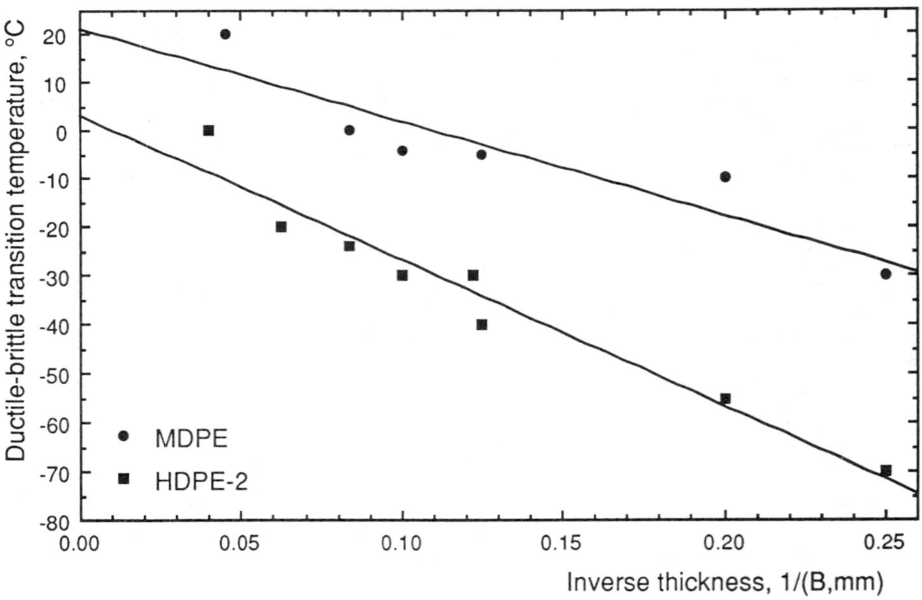

FIG. 10--DBTT *vs.* inverse specimen thickness from Charpy data for MDPE-2 and HDPE-2

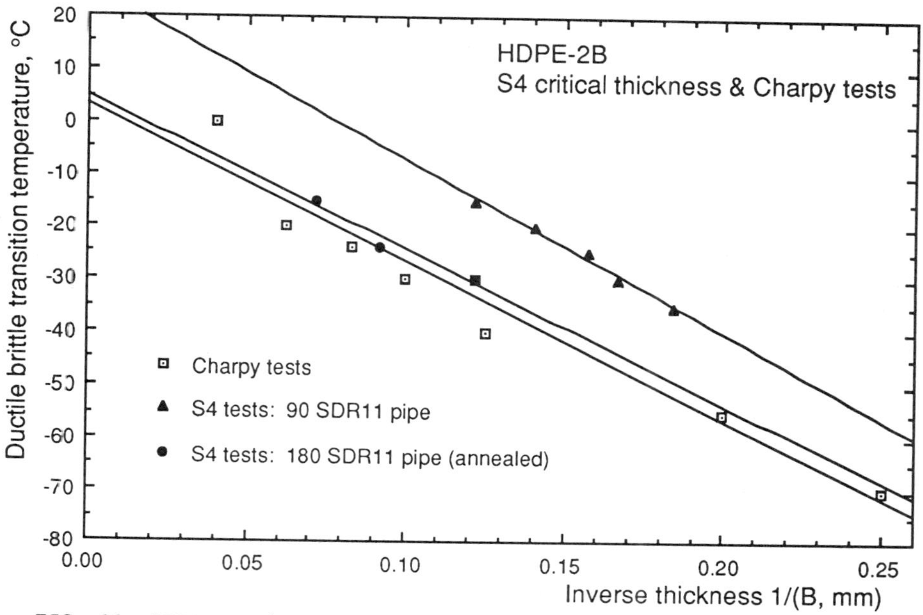

FIG. 11--DBTT *vs.* inverse specimen thickness from Charpy data, and temperature *vs.* critical thickness from S4 tests, for the same HDPE-2 batch

CONCLUSIONS

RCP failure of pressurised PE pipelines is very rare, but it is too dangerous to be ignored.

The plane strain dynamic fracture toughness of new 'third generation' resins is probably no more than 50% higher than that of standard medium density grades, despite their markedly superior resistance to RCP in full-scale buried pipe tests.

For this and other reasons, the prediction of critical pressure from plane strain fracture toughness, using fracture mechanics analysis, may be unreliable – especially for gas rather than water pressurised pipelines.

The material dependence of the critical pressure above which RCP will self arrest is primarily restricted, at least for polyethylene, to the temperature at which there is an abrupt transition to very high fracture resistance.

Third generation resins score their RCP advantage over medium density grades by having transition temperatures below rather than above their minimum service temperature (usually 0°C).

The transition temperature is strongly thickness dependent, especially in the tougher materials. The use of these materials for thicker pipe of larger diameters running at higher pressures must therefore be backed up by a secure procedure for evaluating the transition temperature and its thickness dependence.

Instrumented Charpy tests on sharp-notched specimens across a range of temperatures may have an important role to play in such a procedure.

ACKNOWLEDGEMENTS

The authors wish to express their appreciation of a Special Research Grant for this project from the UK Science and Engineering Research Council. Robert Morgan wishes to acknowledge additional support from Uponor Aldyl (UK) and from Neste Polyeten (Sweden). The tests on water pressurised pipe were carried out by Christopher Greenshields, who is supported by the UK Water Research Centre.

REFERENCES

[1] Greig, J. M., Leevers, P. S. and Yayla, P., "Rapid Crack Propagation in Pressurised Plastic Pipe. I: Full Scale and Small Scale RCP Testing", Engineering Fracture Mechanics Vol. 42, No. 4, 1992, pp 663-673.

[2] Greig, J. M., "Rapid Crack Propagation in Hydrostatically Pressurised 250 mm Polyethylene Pipe", Proceedings of the Seventh International Conference on Plastic Pipes (Bath, England, 19-22 Sept. 1988). Plastics and Rubber Institute, London, 1988.

[3] Yayla, P. and Leevers, P. S., "Rapid Crack Propagation in Pressurised Plastic Pipe. II: Critical Pressures for Polyethylene Pipe", Engineering Fracture Mechanics Vol. 42, No. 4, 1992, pp 675-682.

[4] Wheel, M. A. and Leevers, P. S., "High Speed Double Torsion Tests on Tough Polymers. I: Linear Elastic Steady State and Dynamic Analysis". To be published in International Journal of Fracture, 1993.

[5] Wheel, M. A. and Leevers, P. S., "High Speed Double Torsion tests on tough polymers. II: Nonlinear elastic dynamic analysis". To be published in International Journal of Fracture, 1993.

[6] Leevers, P. S. and Yayla, P., "Thickness Effects on Rapid Crack Propagation in Polyethylene Pipe". Proceedings of the Twelfth Plastic Fuel Gas Pipe Symposium (Boston, Massachusetts, September 24-26 1991). American Gas Association, Arlington VA, 1991.

[7] O'Donoghue, P. E., Green, S. T., Kanninen M. F., and Bowles, P. K., "The Development of a Fluid/Structure Interaction Model for Flawed Fluid Containment Boundaries with Applications to Gas Transmission and Distribution Piping", Computers and Structures Vol. 38 No. 5/6, 1991, pp 501-513.

[8] Kanninen, M. F., Leung, C. P. and O'Donaghue, P. E., "Procedures to Prevent Rapid Crack Propagation in PE Gas Pipes", Proceedings of the Eighth International Conference on Plastic Pipes (Koningshof, Netherlands, 21-24 September 1992). Plastics and Rubber Institute, London, 1992.

Phillip A. Sharff[1] and Steven J. DelloRusso[1]

EFFECTS OF ACID ENVIRONMENT AND CONSTANT DEFLECTION ON PVC SEWER PIPE

REFERENCE: Sharff, P.A. and DelloRusso, S.J., "**Effects of Acid Environment and Constant Deflection on PVC Sewer Pipe,**" Buried Plastic Pipe Technology: 2nd Volume, ASTM STP 1222, Dave Eckstein, Ed., American Society for Testing and Materials, Philadelphia, 1994.

ABSTRACT: A test study was undertaken to determine the effects of a sustained exposure to sulphuric acid solution on pipe stiffness of PVC sewer pipes held at fixed 5% deflection. This study was part of a multi-participant round-robin program to evaluate and develop a test method for long-term effects of deflection in various environments on plastic sewer pipe. Sections of 6 in. (152.4 mm) diameter SDR 35 PVC sewer pipe ASTM D 3034 - Specification for Type PSM Poly (Vinyl Chloride) (PVC) Sewer Pipe and Fittings 6 in. (152.4 mm) long were tested for pipe stiffness by the parallel plate method ASTM D 2412 - Test Method for Determination of External Loading Characteristics of Plastic Pipe by Parallel-Plate Loading, as received from three suppliers. Pipe specimens were then immersed in a 1.0N sulphuric acid solution for 1 week and retested for pipe stiffness. The pipe sections were then held at 5% deflection in special test fixtures and continuously immersed in a 1.0N solution of sulphuric acid at room temperature. The pipe stiffness of each specimen was measured at various time intervals over a period of two years, without removing them from the acid bath. Results were analyzed to evaluate the effects of the acid immersion, coupled with the fixed deflection. One control pipe specimen was fixed at 5% deflection in an air-only environment and tested for pipe stiffness at the same time intervals as the acid-immersed specimens.

Test data indicates a minimal effect of fixed deflection and acid immersion over time on short-term pipe stiffness. Analysis of load data illustrates the significant relaxation response of PVC pipe under constant deflection, i.e. the decrease in wall stress over time. Two year data exhibit good correlations to the design-oriented viscoelastic predictions derived from Findley's power law. The influence of filler content on stiffness behavior is examined but no conclusive relationship is apparent. The influence of the test fixture on interpretation of results is discussed and recommendations for improvements are presented.

KEY WORDS: Poly (vinyl chloride) (PVC) pipe, sewer pipe, sewer acid, pipe stiffness, stress relaxation, viscoelastic behavior.

---

[1] Staff Engineer and Senior Engineer, respectively, Simpson Gumpertz & Heger Inc., 297 Broadway, Arlington, MA 02174

INTRODUCTION

The use of thermoplastic pipe for buried sewers and other gravity-flow applications continues to expand. Among other attributes, thermoplastic pipe is recognized as having good resistance to aggressive environments; however, industry currently has no standard test method for measuring the relative long-term effects of environmental agents combined with sustained stress (or strain) on its stiffness properties. Buried sewer pipe is frequently expected to remain in service 50 years or longer, sometimes subjected to aggressive environments such as sewer acids; thus, long-term properties under service conditions are a critical design consideration.

In 1990 a test program was initiated in which specimens of 6 in. (152.4 mm) diameter poly (vinyl chloride) (PVC) pipe were clamped at a fixed 5% deflection (decrease in vertical diameter) and immersed in a 1.0N solution of sulphuric acid at room temperature. A control specimen was clamped at the same deflection in air only. Specimens were monitored for stiffness properties at periodic intervals. These tests were part of a larger round-robin test program to investigate a standard method for determining the effects on the stiffness properties of thermoplastic pipe held at constant strain and exposed to various environments for long periods of time. Similar tests conducted earlier at Utah State University are reported by Bishop [1]. Requirements of the round-robin test included results on short-term pipe stiffness properties up to one year of exposure. The results reported herein extend beyond the round-robin test requirements and include exposure periods up to two years and analysis of data for long-term behavior of pipe specimens.

A very beneficial aspect of extended studies such as conducted here is the quantifiable information obtained on the long-term stiffness properties of the pipe material. It is essential that the pipe designer consider the time-dependent behavior of buried plastic pipe so that reasonable predictions can be made of stresses, strains, and stability under sustained loading conditions. In order to accomplish this, appropriate means for modeling the viscoelastic response of the material must be available (see Ref. [2] for a thorough discussion of viscoelasticity). Various design approaches reported in the literature [2,3,4,5] take into account the viscoelastic behavior of the plastic pipe material. The American Association of State Highway and Transportation Officials (AASHTO) Standard Specifications for Highway Bridges [6] requires the use of an effective long-term elastic modulus in buckling equations to account for the viscoelastic behavior of thermoplastic pipe. Unfortunately, sparse data is available on long-term stiffness properties of the many commercially available thermoplastic pipe materials. The test method and analyses used in this study provide an efficient means of obtaining such properties.

This paper discusses the significance of test findings relative to the following:

- Effects of sulphuric acid exposure and fixed deflection on pipe stiffness
- Time-dependent relaxation response
- Viscoelastic properties for design
- Influence of test configuration on interpretation of results

TEST PROGRAM

The test procedures used in this study were established as part of an overall round-robin program initiated by an ASTM Task Group under the auspices of ASTM Committee F-17 on Plastic Piping Systems. The round-robin test procedures are provided in Ref. [1]. Details of this study, including work extended beyond the requirements of the round-robin program are given below.

Specimens

A total of 13 pipe specimens were tested in this study - 12 acid-immersion specimens and one air-only control specimen. Nine of the acid-immersion specimens were cut from pipe provided, as part of the round-robin study, from two manufacturers (Manufacturer I and II). Three additional acid-immersion specimens and the one air-only specimen were cut from pipe from a third manufacturer (Manufacturer III). This pipe was purchased locally and was not part of the round-robin study. All pipe were ASTM D 3034 SDR 35 PVC sewer pipe, 6-inch (152.4 mm) nominal diameter.

Pipe specimens were cut to nominal 6-inch (152.4 mm) lengths. Specimen length, wall thickness and outside diameter (OD) dimensions were measured and recorded as described in ASTM D 2122 - Method for Determining Dimensions of Thermoplastic Pipe and Fittings. The average inside diameter (ID) was determined by subtracting twice the average wall thickness from the average OD. All computed percent deflections are based on this value of ID.

Table 1 provides initial data on test specimens.

Table 1 - Initial Data on Test Specimens

Manufacturer	I	II	III
No. of Specimens	6	3	4
Cell Class ASTM D 1784	Not given	Not given	12364C
Specified Modulus of Elasticity psi (kPa) ASTM D 1784	-	-	400,000 (2,800,000)
Average Wall Thickness, in. (mm)	0.201 (5.10)	0.188 (4.78)	0.196 (4.98)
Average Inside Diameter, in. (mm)	5.877 (149.3)	5.882 (149.4)	5.871 (149.1)
Ash Content (%) ASTM D 229	3.9	4.5	11.7

Test Fixtures

Fixed-deflection test fixtures for acid-immersion specimens were fabricated from stainless steel because of the corrosive test environment. Dimensions of fixture parts specified for the study are provided in Figure 1. An assembled test fixture is shown in Figure 2. This set-up

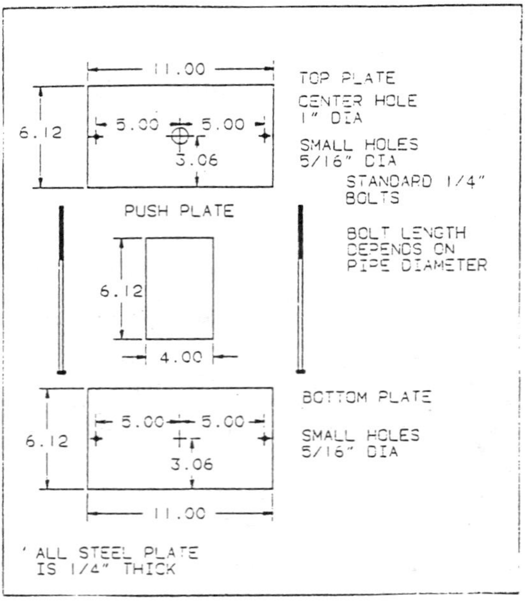

Figure 1 - Dimensions of Fixed Deflection Test Fixture for Acid Immersion Specimens. (1 inch = 25.4 mm)

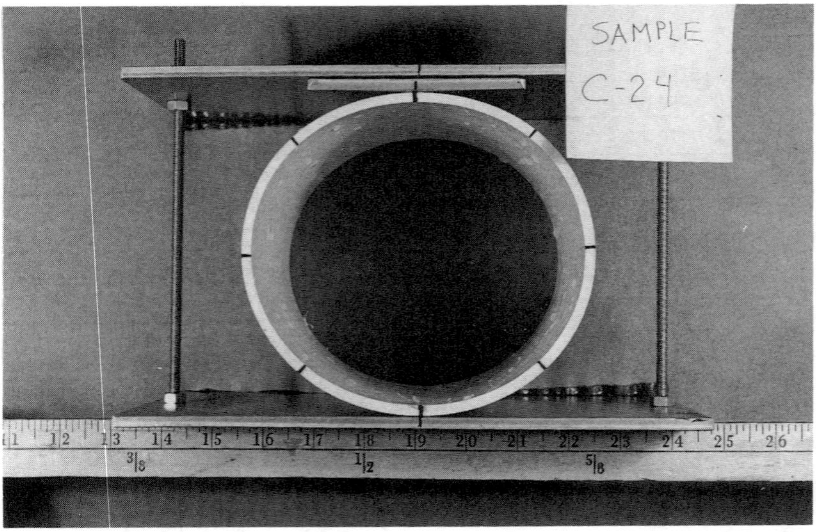

Figure 2 - Fixed Deflection Test Fixture for Acid Immersion Specimen.

permitted additional incremental deflection of the pipe by inserting the load head through the top plate and bearing on the push plate. The air-only specimen was clamped at fixed deflection using steel channels, 10x2-1/2x1/4 in. (254 x63.5 x 6.4 mm) for top and bottom plates and 6x2x3/16 in. (152.4 50.8 x 4.8 mm) for the push plate.

A specimen containment system was selected that permitted continual acid immersion even during load testing. Each test specimen, except for the control specimen in air, was immersed in an acid environment in separate plastic containment vessels, which could be mounted in the Zwick 1474 Universal Testing machine, thus eliminating the necessity of removing the specimen from the test environment to measure pipe stiffness (Figure 3).

Test Procedures

An initial pipe stiffness test to about 10% deflection was conducted for each specimen prior to immersion using the parallel plate method (ASTM D 2412). This enabled determination of initial short-term pipe stiffness between 5% and 7.5% deflection (PS_o) as shown in Figure 4a. All test specimens, except for the air-only control specimen, were then immersed, undeflected, in a 1.0N sulphuric acid (H_2SO_4) solution for one week. After the one week immersion, a second pipe stiffness test to 10% deflection was conducted for each of the immersed specimens. Each specimen was then placed in the fixed deflection test fixture and clamped at 5% deflection and, except for the air-only specimen, immersed in the acid bath. Periodic titration tests and corrections were made to maintain acid solution normality.

A short-term pipe stiffness test from the deflected 5% condition to about 10% deflection was conducted for each immersion specimen and the air-only specimen after the following immersion time intervals (as required by the initial round-robin study): 24 hours, 1, 2, and 4 weeks and at 1 month increments thereafter for 1 year. This enabled determination of short-term pipe stiffness between 5% and 7.5% deflection at each time interval (PS_t) as shown in Figure 4b. Additional pipe stiffness tests were conducted at 18 months and 27 months. All testing was conducted at room temperature, 73°F (23°C).

RESULTS AND DISCUSSION

Short-Term Pipe Stiffness

Figures 5 and 6 illustrate the change in incremental short-term pipe stiffness of test specimens when deflection is increased under a short-term load from the fixed 5% deflection to a new deflection of 7.5% at periodic test intervals up to 27 months. Figure 5 shows mean results for acid-immersed specimens for Manufacturers I, II, III and Figure 6 compares mean results of the acid-immersed specimens from Manufacturer III with the air-only specimen from Manufacturer III. Schematic representation of terms is provided in Figure 4.

It is apparent from Figure 5 that fixed deflection and extended immersion in 1.0N sulphuric acid has not significantly affected the short-term pipe stiffness of specimens held at 5% deflection. Variations are on the order of 15% or less of the initial pipe stiffness and show no relationship to time of exposure. Manufacturer II and III specimens show an almost immediate increase in short-term pipe stiffness of about 10 to 15%, and then later fluctuations down and up within the 15% range.

Figure 3 - Load Testing of Continuously Immersed Specimen in Individual Containment Vessel.

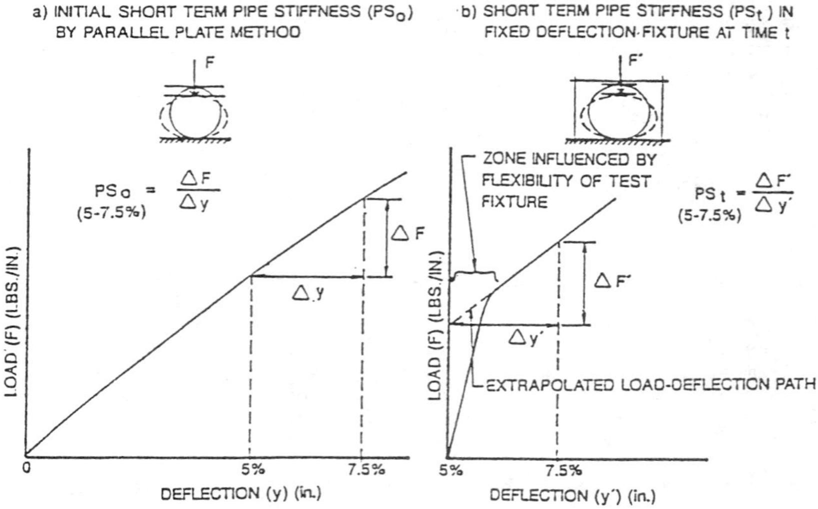

Figure 4 - Typical Load-Deflection Plots of PVC Pipe Specimens for a) Initial Short Term Pipe Stiffness and b) Short Term Pipe Stiffness at Time t.

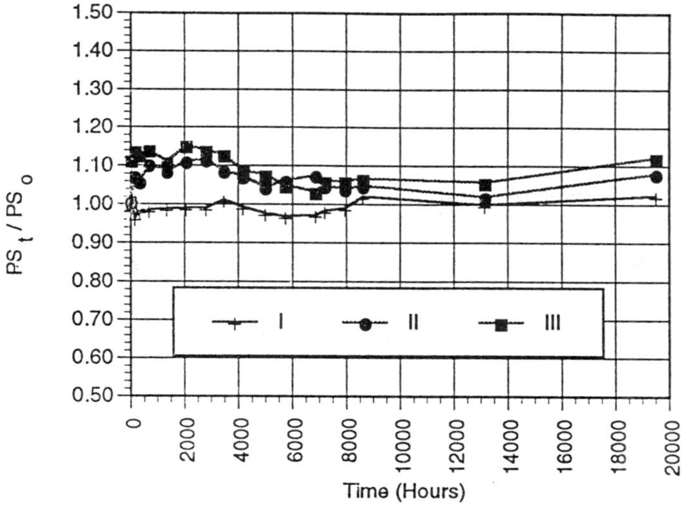

Figure 5 - Short Term Pipe Stiffness (5% to 7.5% Deflection) of PVC Pipe Held at 5% Deflection and Immersed in 1N Sulphuric Acid

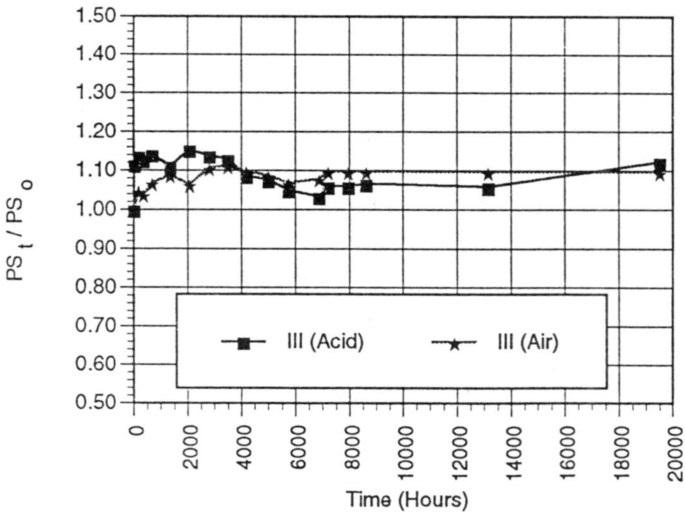

Figure 6 - Short Term Pipe Stiffness (5% to 7.5% Deflection) of PVC Pipe Held at 5% Deflection: Air vs. Immersed in 1N Sulphuric Acid

Manufacturer I specimens show essentially no change in short-term pipe stiffness over the test period.

Examination of Figure 6 indicates no marked distinction in the short-term stiffness response of the acid-immersed pipe specimens compared to that of the control specimen exposed to air only. These results are similar to those obtained by Bishop [1].

The short-term pipe stiffness is useful in characterizing the deformation response of the buried pipe subjected to loads where the short-term response is of prime interest. This would be the case, for example, under quasi-instantaneous loadings such as traffic live loads or dead weight surcharges after original soil/pipe installation has stabilized. Under such loadings, the deformation behavior of the pipe can be reasonably determined using initial stiffness properties as determined by the parallel plate method.

The slight variations in response shown in Figures 5 and 6 indicate that filler quantities, as indicated by ash content (Table I) may play some role in the retention of short-term pipe stiffness. Pipe specimens with a higher ash content (Manufacturer II at 4.5% and Manufacturer III at 11%) show a slight increase in short-term pipe stiffness, whereas specimens with lower ash (Manufacturer I at 3.9%) show almost no change in short-term stiffness over the test period. The mechanism of influence of filler content on variations in pipe stiffness over time is not well established and further controlled study appears warranted.

Time-Dependent Stress Relaxation

Figure 7 tracks the decay in load required to sustain the 5% fixed deflection for all acid-immersed specimens. The decay in load over time is a measure of the time-dependent stress relaxation in the pipe material. Relaxation is defined as the decrease in stress with time under constant strain. Greater stress relaxation is exhibited for pipe specimens from manufacturer I and III than for manufacturer II specimens. After 27 months (19,500 hrs) of testing Manufacturer I and III specimens have relaxed to about 45% of initial stress, whereas Manufacturer II specimens have relaxed to about 60% of initial stress.

Comparison of the relaxation response of acid-immersed specimens to the air-only specimen from the same pipe (III) is shown in Figure 8. Almost from the beginning, acid-immersed specimens exhibit about 10% less load retention (i.e. greater relaxation) at 5% deflection; this difference remains about constant over the two years of testing. The differences in deflection fixtures for the air-only and the acid-immersed specimens may have had some influence on the different responses.

Note that for all test specimens, stress relaxation is most pronounced within the first month (700 hrs) of testing. After about 10 months (7000 hrs), relaxation begins to level out, but continues at a slow rate even at 27 months (19,500 hrs).

The time-dependent relaxation behavior exhibited in test specimens should approximate that of a pipe in the buried condition if one assumes that the buried pipe is in a state of fixed deflection (constant strain). Some investigators adopt the concept of fixed deflection in the analysis of long-term behavior of buried thermoplastic pipe. It should be recognized, however, that early in the life of the soil/pipe system, pipe deflections can and do increase, largely due to time-dependent deformation of the surrounding soil. In this condition the

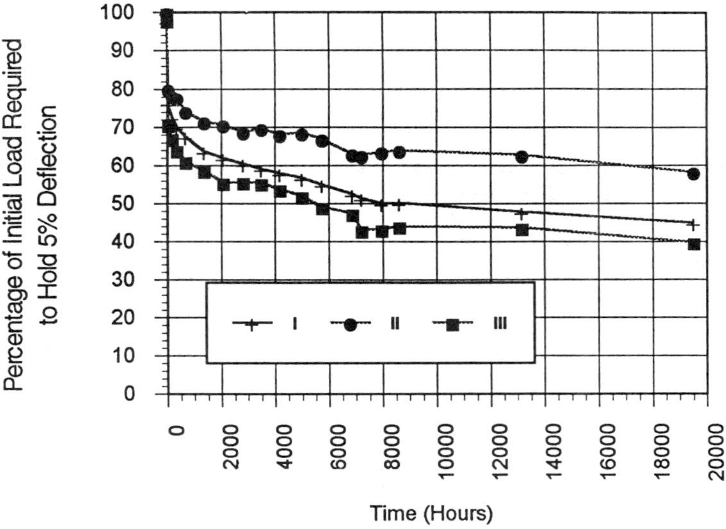

Figure 7 - Relaxation Response of PVC Pipe Held at 5 % Deflection Immersed in 1N Sulphuric Acid

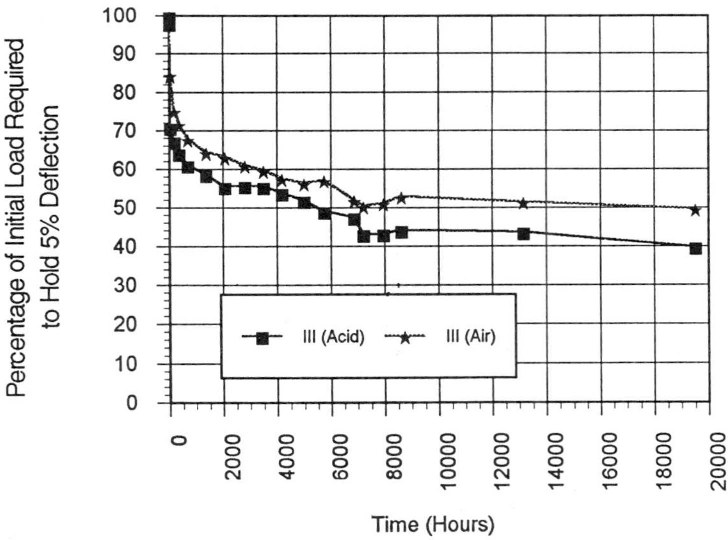

Figure 8 - Relaxation Response of PVC Pipe Held at 5 % Deflection: Air vs Immersed in 1N Sulphuric Acid

time-dependent response of the pipe is termed creep; that is, the increase in strain (deflection) under constant stress. As discussed below, both relaxation and creep behavior can be quantified in terms of the viscoelastic modulus.

Viscoelastic Behavior

Data from time-dependent relaxation response of pipe samples are applicable to a generalized viscoelastic expression for the PVC pipe material. The approach employed here is derived from Findley's power law for creep [7], further developed and simplified for design applications in terms of modulus as follows [2]:

$$1/E_v = 1/E_o + t^n/E_t \qquad \text{Eq. 1}$$

where:
- E_v = time-dependent viscoelastic modulus, psi (kPa)
- E_o = elastic time-independent modulus, psi (kPa)
- E_t = constant which characterizes the time-dependent stiffness
- t = time after loading (hrs.)
- n = dimensionless material constant which characterizes time-dependent behavior

In the analysis of test data, E_o, the initial flexural modulus, is taken as the tangent slope of the stress strain plot from the parallel plate test on specimens after one-week conditioning in the test environment before clamping in the fixed deflection test fixture. E_v is computed for each test time (t) as the flexural secant modulus at 5% deflection. Flexural stresses and strains, derived from Roark's solution for a circular ring under a concentrated load [8] are calculated as follows:

$$\sigma = 0.955\ PD/h^2 \qquad \text{Eq. 2}$$

$$\epsilon = 4.27\ (\Delta y/D)\ (h/D) \qquad \text{Eq. 3}$$

where:
- σ = stress, psi (N/mm^2)
- ϵ = strain, in./in. (mm/mm)
- P = load, lbs/in. (N/mm)
- D = mean diameter, in. (mm)
- h = thickness, in. (mm)
- Δy = vertical deflection, in. (mm)

Plotting E_v/E_o vs. log time and curve fitting Equation 1 to data yields values for constants E_t and n. Figures 9 and 10 show semi-log plots for E_v/E_o and fitted curves computed from test data for pipe specimens from the different manufacturers.

Table 2 summarizes viscoelastic constants for the test pipe, including the pipe exposed to air only. Also included in the table are constants derived from Findley's tests on rigid PVC in tensile creep begun in 1957 and continued for 26 years. The Findley constants, still cited in design references for PVC [e.g., 2,9], are plotted according to Eq. 1 in Figure 10 for comparative purposes. Note the marked difference between test-derived E_t values and E_t values derived from Findley's tests.

Table 2 - Viscoelastic Constants

Pipe Type	E_o (psi)*	E_t (psi)*	n
I	358,000	3,000,000	0.23
II	464,000	2,800,000	0.16
III	444,000	1,900,000	0.19
III-air	428,000	2,200,000	0.19
Derived from Findley tests [2]	557,000	20,500,000	0.31

*1 psi = 6.894 kPa

Constants E_t and n, along with experimentally determined E_o, are now sufficient to compute the viscoelastic modulus E_v at any time and at any stress for similar environment and temperatures within the linear viscoelastic range. The linear viscoelastic range is that range of deflections for which a linear proportionality exists between stress and strain at any time after either a constant stress (creep) or constant strain (relaxation) is imposed. It has been shown that for PVC, the linear viscoelastic range includes strains up to about 5% deflection [2].

Within the linear viscoelastic range, both creep and relaxation can be approximated by a single viscoelastic modulus, E_v, that is independent of the magnitude of stress or strain. Thus, the method illustrated above for computing E_v from fixed deflection tests provides the pipe analyst with a very practical tool for predicting the stiffness response of commercially available PVC sewer pipe subjected to long-term loads or deflection. Values for E_v so obtained can be used in appropriate elastic design equations (e.g. buckling equations) to characterize time-dependent structural behavior. (The temperature dependence of E_v must be considered when variations in operating temperatures are anticipated).

An obvious benefit of developing the time-dependent relationship given in Equation 1 is the predictive capability of extrapolating viscoelastic curves to longer time periods. As an illustration, extrapolation of curves derived from test data (Figures 9 and 10) to 10 and 50 years yields values for E_v shown in Table 3. These results suggest that long-term viscoelastic moduli may vary significantly from one manufacturing source of another. Further testing of larger sample sizes are needed to determine the validity of these apparent variations.

Table 4 provides extrapolated viscoelastic moduli derived from the Findley test data and 50-year values from the AASHTO design specification [5]. These values represent currently available design values for long-term moduli. Although comparable in magnitude to the test-derived values in Table 3 (assuming negligible effects of acid as seems to be the case), neither the AASHTO nor the Findley values appear to be particularly conservative.

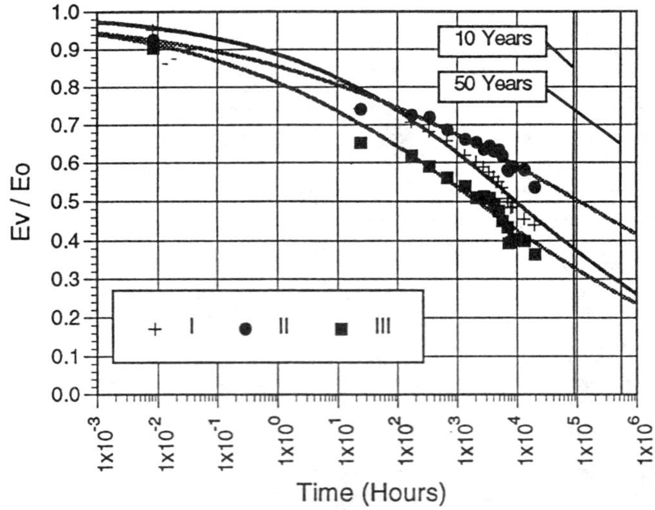

Figure 9 - Viscoelastic Behavior of PVC Pipe at 5% Deflection Immersed in 1N Sulphuric Acid

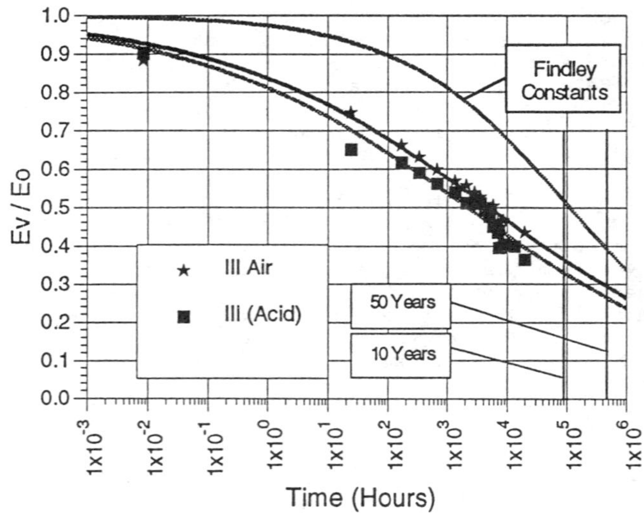

Figure 10 - Viscoelastic Behavior of Air vs. Acid - Immersed PVC Pipe. Also Shown is Curve Derived from Findley Creep Tests on Rigid PVC Tensile Coupons

Table 3 - Extrapolated Values for Viscoelastic Modulus From Tests (psi)*

Test Pipe	Cell Class	E_o	E_v (10 yrs.)	E_v (50 yrs.)
I	-	358,000	136,000	107,000
II	-	464,000	236,000	208,000
III	12364C	444,000	147,000	118,000
III (AIR)	12364C	428,000	156,000	127,000

Table 4 - Available Values for Viscoelastic Modulus (psi)*

Source	Cell Class	E_o	E_v (10 yrs.)	E_v (50 yrs.)
Findley Data	NA	557,000	289,600	217,200
AASHTO	12454C	400,000	NA	140,000
AASHTO	12364C	400,000	NA	158,400

* 1 psi = 6.894 kPa

It would appear beneficial to the design and user community that a standard test method for qualifying the long-term performance of PVC sewer pipe include a methodology such as outlined herein for establishing values for long-term viscoelastic moduli.

Performance of Test Fixture

Overall, the text fixture and procedures employed in this program provided useful and consistent results. The use of individual acid containment vessels and in situ load testing of immersed specimens minimized the hazards associated with repeated removal, washing, and re-immersing specimens. However, several anomalies were noted during testing which may lead to small, but not insignificant, measurement errors. These are:

- **Excessive flexibility** - Flexibility in the fixed deflection test fixture resulted in a sizeable deflection offset in the initial load-deflection curves for incremental short-term stiffness tests of acid-immersed specimens. The offset is illustrated in Figure 4b. Extrapolation of the load-deflection curve through the offset region reduces accuracy in measuring load at 5% deflection. This offset zone was greatly reduced in the air-only test fixture using stiffened channels for top and bottom plates, indicating the offset is caused primarily by bending of the top and bottom plates specified in Figure 1 for immersion specimens.

- **Movement of pipe** - Progressive rotation of acid-immersed pipe specimens off the load center occurred with successive load testing in the test fixture. This probably resulted from a slight eccentricity in contact of the load head. In situ testing in acid containers precludes monitoring this effect and presents one argument against this type of arrangement unless frequent checks and adjustments to pipe position are incorporated into the test protocol.

- **Load eccentricity** - Slight eccentricities in contact of the loading nose with the steel push plate caused rotation of the push plate before complete release from the top plate. This effect manifested as "cupping" of the load-deflection curve and adds to measurement errors associated with curve extrapolation, similar (and additive) to those caused by excessive flexibility of fixture plates.

Recommendations to be considered in developing a standard test method include:

- Provide stiffer top and bottom test plates to minimize effects of plate bending; channel sections are most efficient.
- Conduct preliminary load tests on pipe in the immersion fixture to verify concentric contact of the push plate on pipe specimen, and provide permanent gage marks to assure proper alignment is maintained.
- If individual containments vessels are used, make periodic checks of pipe specimens to inspect for and correct progressive rotation of the pipe specimen in the test fixture.

CONCLUSIONS

This test study leads to the following conclusions:

- The effect of 1.0N sulphuric acid immersion combined with fixed deflection on the short-term pipe stiffness of the PVC pipe specimens appears to be minimal.
- Significant relaxation of stress in the pipe occurs over the two year period at a fixed 5% deflection. Marked differences in relaxation behavior of pipe from different sources was observed.
- The data obtained by this long-term test program provides very useful design information on the viscoelastic behavior of commercially available PVC pipe. A method, derived from Findley's power law, for determining viscoelastic modulus (E_v) after any time of sustained stress or strain is directly applicable to tests results.
- Improvements in the test fixture and protocol are recommended to minimize possible errors interpreting test results.

REFERENCES

[1] Bishop, R.R., "Retention of Pipe Stiffness for Polyvinyl Chloride (PVC) Pipe Samples Exposed to Various Environments and Constant Strain," ASTM STP 1093, Buried Plastic Pipe Technology, Buczala, G.S. and Cassady, M.J., editors, American Society for Testing and Materials, Philadelphia, PA, 1990.

[2] American Society of Civil Engineers, ASCE Manuals and Reports on Engineering Practice No. 63, Structural Plastics Design Manual, American Society of Civil Engineers, New York, NY, 1984.

[3] Chambers, R.E. and Rund, R.E., "Structural Design of Pipe via AASHTO Specifications," Second Conference on Structural Per-

formance of Pipes, Sargand, S.M., Mitchell, G.F., Hurd, J.O., editors, Columbus, Ohio, March 1993.

[4] Chua, K.M and Lytton, R.L., "Viscoelastic Approach to Modeling Performance of Buried Pipes," Journal of Transportation Engineering, Vol. 115, No. 3, May 1989, American Society of Civil Engineers, New York, NY.

[5] Taprogge, Rainer H., "Large Diameter Polyethylene Profile-Wall Pipes in Sewer Application", Proceedings of the International Conference on Underground Plastic Pipe, B. Jay Schrock, Editor, American Society of Civil Engineers, New York, NY, 1981. p. 175-190.

[6] American Association of State Highway and Transportation Officials, "Standard Specifications for Highway Bridges," 15th Edition, Section 18, Soil-Thermoplastic Pipe Interaction Systems, Washington, D.C., 1992.

[7] Findley, W.N., "26-Year Creep and Recovery of Poly(Vinyl Chloride) and Polyethylene, Polymer Engineering and Science, Volume 27, No. 8, April 1987.

[8] Roark, Raymond J., Formulas for Stress and Strain, Third Edition, McGraw-Hill Book Company, New York, NY, 1954, p. 156.

[9] Throne, J.L. and Progelhof, R.C., "Creep and Stress Relaxation in Engineering Plastics, Vol. 2," Engineered Materials Handbook, ASM International, Metals Park, Ohio, 1988, pp. 659 to 678.

Trenchless Construction

Thomas W. Hawkins[1] and Thomas R. Mass[2]

THE EFFECTS OF SULFURIC ACID ON CALCIUM CARBONATE FILLED PVC SEWER PIPE COMPOUNDS

REFERENCE: Hawkins, T. W. and Mass, T. R. **"The Effects of Sulfuric Acid on Calcium Carbonate Filled PVC Sewer Pipe Compounds"**, Buried Plastic Pipe Technology: 2nd Volume, ASTM STP1222, Dave Eckstein, Ed., American Society for Testing and Materials, Philadelphia, 1994.

ABSTRACT: Even though calcium carbonate filled rigid polyvinyl chloride (UPVC) sanitary sewer pipe has been used since 1973, concerns are occasionally raised regarding its possible corrosion by sulfuric acid present in the sewer gas environment. In this study two calcium carbonate filled PVC pipe compounds, under prestressed conditions, are exposed to sulfuric acid (20%) for extended periods. The effects of the sulfuric acid on the calcium carbonate in the compounds are qualitatively measured by scanning electron microscopy (SEM) and quantitatively determined by wavelength dispersive x-ray microanalysis (WDS). The study shows that only the calcium carbonate immediately at the surface is affected by the sulfuric acid. The calcium carbonate below the surface is well dispersed and completely encapsulated in a continuum of PVC with no inner-connectivity between neighboring calcium carbonate particles and is unaffected. Results for samples exposed for 14 days, 30 days, 90 days and 6 months are included.

KEYWORDS: Calcium Carbonate, filled PVC compound, corrosion, sulfuric acid aging, sewer pipe, microscopy

---

INTRODUCTION

In the 1960's, rigid polyvinyl chloride (UPVC) was introduced as pipe in pressure applications [1]. The

---

[1]Director of Material Engineering, The Lamson & Sessions Co., Lamson Vylon Pipe Division, Cleveland, OH 44122.

[2]Senior Plastics Engineer, The Lamson & Sessions Co., Lamson Vylon Pipe Division, Cleveland OH 44122.

compounds used had an ASTM cell classification of 12454B or C [2]. Even these early compounds contained 3 to 5 phr of a fine, approximately one micron average particle size, stearate coated calcium carbonate ($CaCO_3$). The primary purpose of the $CaCO_3$ was to scour the extrusion tooling.

When UPVC was expanded into non-pressure applications, such as electrical conduit and duct, and sanitary sewer and drain pipe, new lower-tensile compounds having significantly higher levels of $CaCO_3$ were introduced. In many applications, courser grades of uncoated $CaCO_3$ were used.

In sewer pipe applications, two cell classifications of filled UPVC compound are specified, 12364 and 12164 (with chemical resistance rating of A, B or C) [3-7]. The 12364 cell classification can be made with up to 30 parts per hundred (phr) of $CaCO_3$ and 12164 with up to 60 phr $CaCO_3$. For sewer pipe applications, $CaCO_3$ is primarily used to raise the modulus of elasticity which results in increased pipe stiffness, an important flexible sewer pipe property. In some ASTM Standards [5-6] wall thicknesses are based on the materials modulus, i.e., higher modulus materials allow for thinner walls. Higher filled UPVC sewer pipe has been in the field since the early 1970's with good performance. The higher filled pipes are required to meet the same physical requirements as lower filled pipe. These requirements include resistance to sulfuric acid (H_2SO_4) as well as other chemicals [2, 8, 9].

Yet with this background and performance history, questions and concerns continue to be voiced by both users and competitors regarding the filled sewer pipe corrosion resistance to H_2SO_4 generated in a sewer environment. Certainly a portion of these concerns arise from the documented poor corrosion resistance of concrete pipe [8] caused by the H_2SO_4 reacting with the $CaCO_3$ in the concrete. The chemical reaction taking place being:

$$CaCO_3 + H_2SO_4 \rightarrow CaSO_4 + CO_2\uparrow + H_2O$$

This study takes a very direct physical and analytical approach to what happens to $CaCO_3$ in higher filled UPVC compounds when exposed to H_2SO_4. The study is setup to test the premise, "that because the dispersed particles of $CaCO_3$ in the compound are completely encapsulated in a continuum of PVC polymer, they are protected from attack by the H_2SO_4". Figure 1a is a conceptual representation of $CaCO_3$ dispersed in PVC polymer, showing no contact between particles. Figure 1b is an actual SEM photomicrograph of a 60 phr $CaCO_3$ filled UPVC compound, which confirms the conceptual representation in Figure 1a.

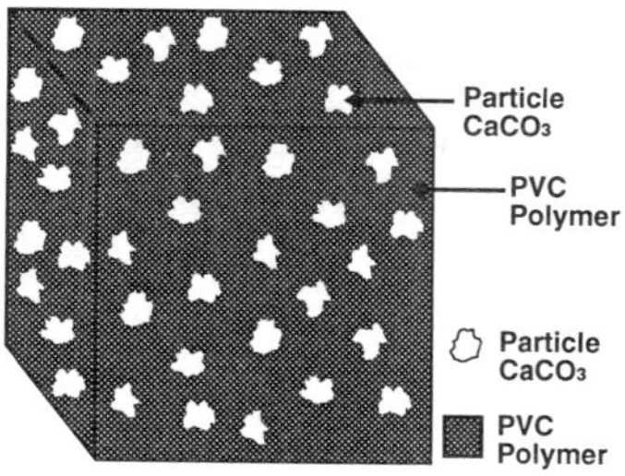

Fig. 1a
Conceptual Representation
of $CaCO_3$ Dispersed in PVC

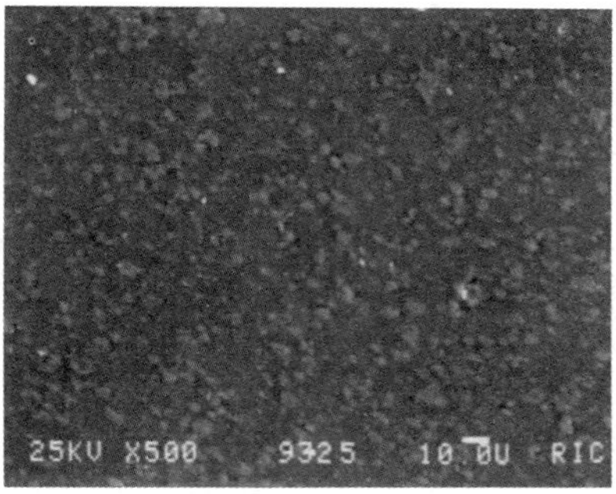

Fig. 1b
Actual SEM Photomicrograph
of 60 phr $CaCO_3$ Filled UPVC

FIG. 1

EXPERIMENTAL

Formulations

Table 1 lists the two UPVC formulations tested in this study. Formulation "A" has an ASTM D-1784 cell class designation of 12364 and Formulation "B" has a cell classification designation of 12164. These compounds are representative of typical UPVC compounds presently used in sewer pipe. An uncoated, coarser grade of $CaCO_3$ was used in this study to simulate what is considered a "worse case" scenario for the H_2SO_4 attack.

TABLE 1--UPVC Formulations Studied

Ingredients	Formulation "A"	Formulation "B"
PVC Resin	100.00 phr	100.00 phr
Stabilizer	0.70	0.70
Lubricants	3.60	4.10
TiO_2	1.00	1.00
$CaCO_3$ (4 micron avg.)	25.00	60.00
ASTM D-1784 Cell Class	12364	12164

Sample Preparation

The test specimens were 1.8mm x 12.7mm x 50 mm (0.070" x 0.500" x 2.000") compression molded strips of Formulation "A" and "B". The test strips were then bonded to a fixture having a radius which created an initial surface fiber stress of 17,237kPa (2500 psi) in the specimens (Figure 2).

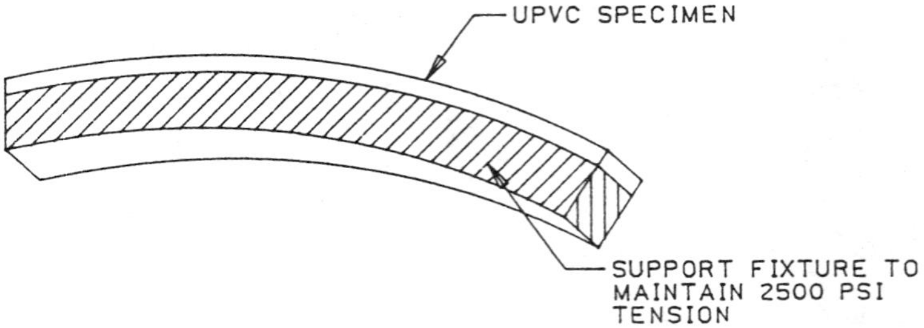

FIG. 2--Schematic of Specimen Mounted to Fixture

Sulfuric Acid Exposure

The stressed UPVC samples were immersed in 20 weight percent concentration H_2SO_4 (pH <0.3). The 20% concentration is taken from the Los Angeles Green Book [10] and, based on a study of sewer lines [8], is considered a relatively aggressive environment. The exposed samples were removed at predetermined intervals (14 days, 30 days, 90 days, etc.) rinsed with deionized water and air dried.

Analytical Measurements

Scanning electron microscopy (SEM) at 500x magnification was used to visibly observe the effect of the H_2SO_4 on the $CaCO_3$ both at the sample surface and five locations deeper into the sample cross section (Figures 3a and 3b).

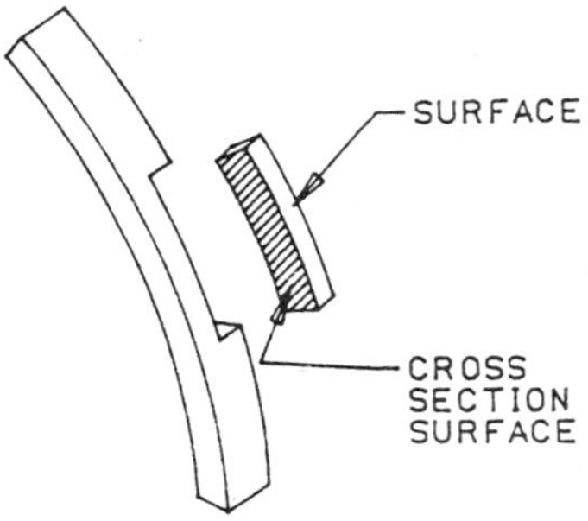

Fig. 3a
Schematic Showing Sample
Surface and Cross Section
Examined After Exposure

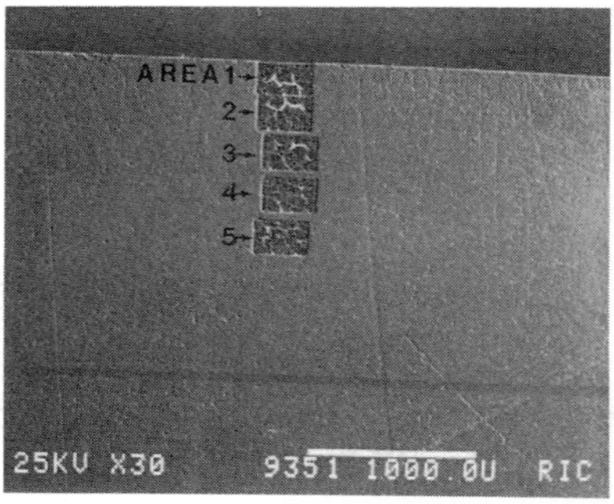

Fig. 3b
30X SEM Photomicrograph
Showing 5 Cross Section
Areas Studied

FIG. 3

Wavelength dispersive x-ray spectroscopy (WDS) microanalysis was used to quantitatively measure the effects of the H_2SO_4 on the $CaCO_3$ at the same locations as the SEM work. The degree of attack on the $CaCO_3$ was measured by looking at the reduction in calcium and the increase in sulfur concentrations after H_2SO_4 immersion.

RESULTS AND DISCUSSION

14 Days Sulfuric Acid Exposure

Figure 4 shows the SEM photomicrographs of the surface of the 25 phr and 60 phr $CaCO_3$ filled UPVC formulations before and after 14 days exposure to 20 weight percent concentration H_2SO_4.

The photomicrographs in Figure 4 show that after 14 days exposure, much of the $CaCO_3$ at the surface of both the 25 and 60 phr filled samples was attacked and removed, leaving voids. Both the 25 phr and 60 phr filled samples showed an equivalent degree of attack on the $CaCO_3$. This would be anticipated because much of the $CaCO_3$ at the surface is not totally encapsulated in the PVC polymer.

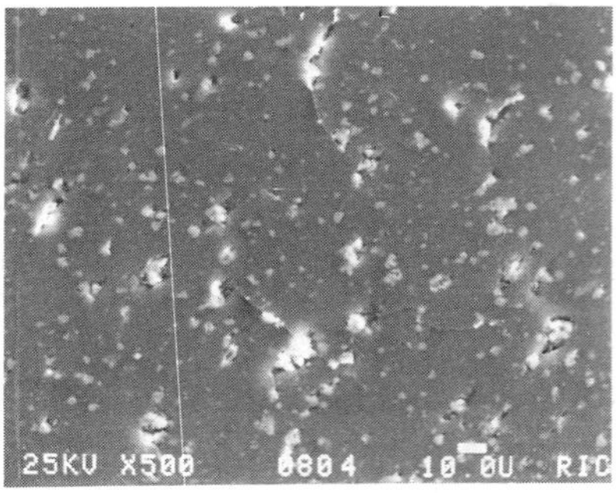

"A" Unexposed

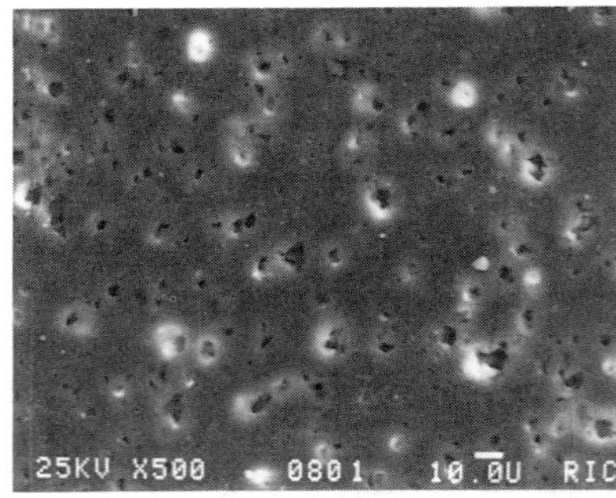

"A" Exposed 14 Days

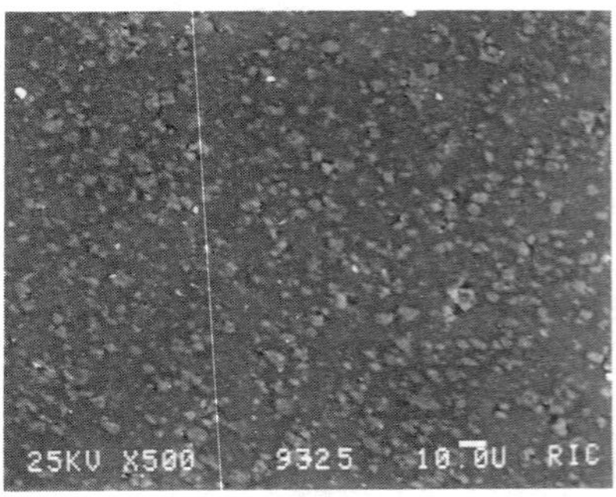

"B" Unexposed

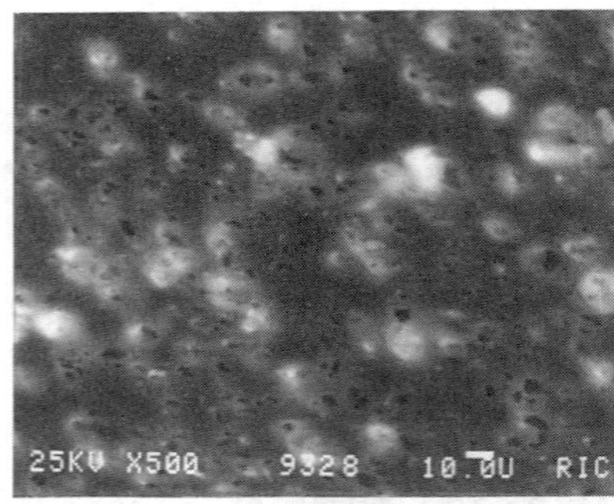

"B" Exposed 14 Days

FIG. 4--500x SEM's of Surface of Formulations "A" and "B" Before and After 14 Days H_2SO_4 Exposure

Figure 5 shows the photomicrographs of cross section Area 2 (0.3mm below surface) of the 25 phr and 60 phr $CaCO_3$ filled UPVC formulations before and after 14 days exposure to the 20% concentration H_2SO_4.

The photomicrographs in Figure 5 show no attack of the $CaCO_3$ in Area 2 for either formulation after 14 days exposure to H_2SO_4. Even though Area 2 is only 0.3mm (0.11") from the surface the H_2SO_4 had no visible effect on the $CaCO_3$. This is consistent with the premise that the $CaCO_3$ is protected from the H_2SO_4 by an envelope of PVC polymer.

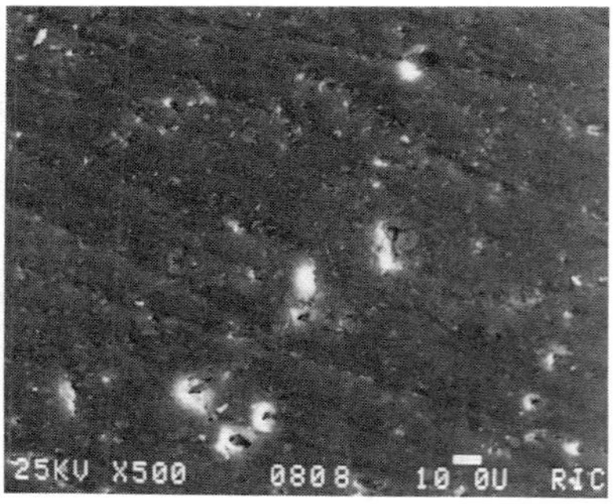

"A" Unexposed

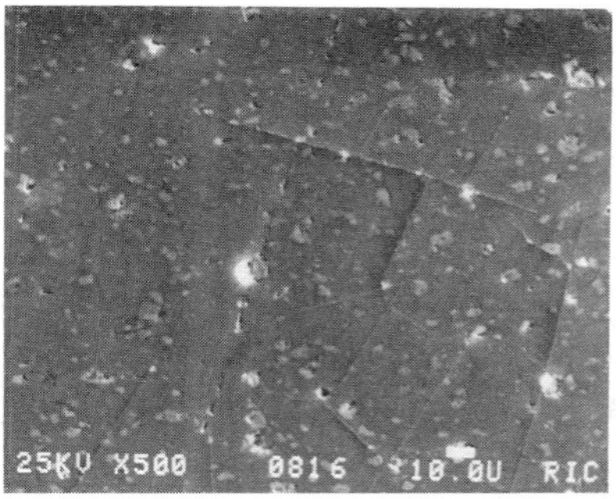

"A" Exposed 14 Days

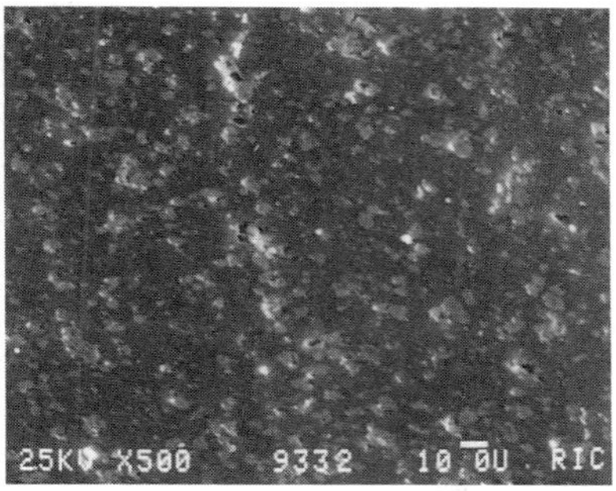

"B" Unexposed

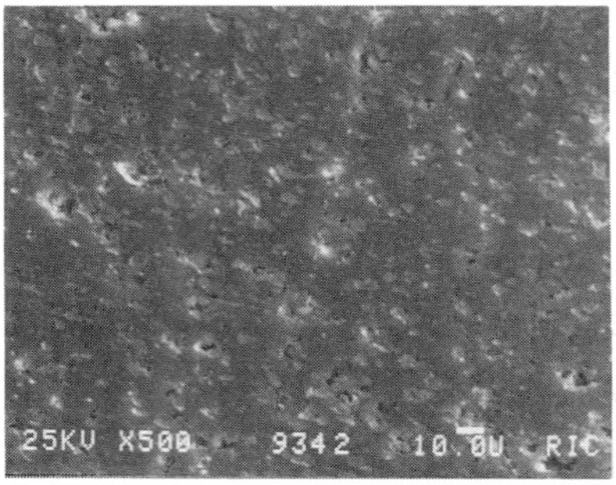

"B" Exposed 14 Days

FIG.5--500x SEM's of the Cross Section Area 2 for Formulations "A" and "B" Before and After 14 Days H_2SO_4 Exposure

In an effort to quantify the effects shown in the SEM work wavelength dispersive x-ray microanalysis (WDS) techniques were used. It is postulated that if H_2SO_4 attacked the $CaCO_3$ there would be a decrease in calcium and an increase in sulfur content. Table 2 shows the results of the WDS analysis of the same surface and cross sectional areas examined in the previous SEM microscopic analysis. Also included in Table 2 are the calcium and sulfur counts for the other four cross sectional areas studied (see Figure 3b).

The WDS data in Table 2 shows there is a significant increase in sulfur and reduction in calcium counts at the surface after 14 days of H_2SO_4 aging. This quantitatively confirms that there is $CaCO_3$ at the surface available for attack by H_2SO_4. Both the 25 phr and 60 phr filled UPVC's showed the same degree of attack.

TABLE 2--Sulfur and Calcium WDS Analysis Results After 14 Days Aging in 20 wt.% Sulfuric Acid

Area Analyzed	Distance Below Surface[1]	Aged	WDS Sulfur Counts		WDS Calcium Counts	
			25 phr	60 phr	25 phr	60 phr
Surface	---	No	7	6	2998	5619
Surface	---	Yes	12	19	1895	2739
Cross Section	0.11mm (0.004")	No	8	5	2586	4713
Area 1	0.11mm (0.004")	Yes	8	9	2600	4565
Area 2	0.03mm (0.012")	Yes	8	6	3039	4932
Area 3	0.56mm (0.022")	Yes	8	4	3140	4793
Area 4	0.83mm (0.033")	Yes	8	5	3078	4776
Area 5	1.12mm (0.044")	Yes	9	4	3621	4893

[1]Mid-Point of Cross Sectional Area

The WDS results for the cross section areas 1 through 5, below the surface, show no attack on the $CaCO_3$. The variation in the sulfur and calcium counts are within the normal accuracy of the experimental procedure and capabilities of the instrumentation.

For the 14-day H_2SO_4 exposure, the SEM and WDS analysis shows that only the $CaCO_3$ immediately at the surface is affected by the H_2SO_4. The work shows that the H_2SO_4 does not penetrate the PVC polymer matrix, therefore the $CaCO_3$ below is unaffected. This portion of the study showed that the 25 phr and 60 phr $CaCO_3$ filled UPVC were attacked to an equal extent and therefore the 25 phr filled compound was dropped from the extended aging work. The 60 phr filled UPVC was selected because it represented a worse case scenario for potential H_2SO_4 attack.

Extended Sulfuric Acid Exposure

In an effort to explore the longer term affects of the H_2SO_4 on $CaCO_3$ filled UPVC the exposure work was continued beyond the initial 14-day aging. In this paper, 30 and 90 day exposure results on 60 phr $CaCO_3$ filled UPVC are presented. Longer term exposure work continues and will be reported as available.

Figure 6 shows SEM photomicrographs of the surface of the 60 phr $CaCO_3$ filled specimen before and after 14, 30 and 90-day exposure to 20 weight percent concentration of H_2SO_4.

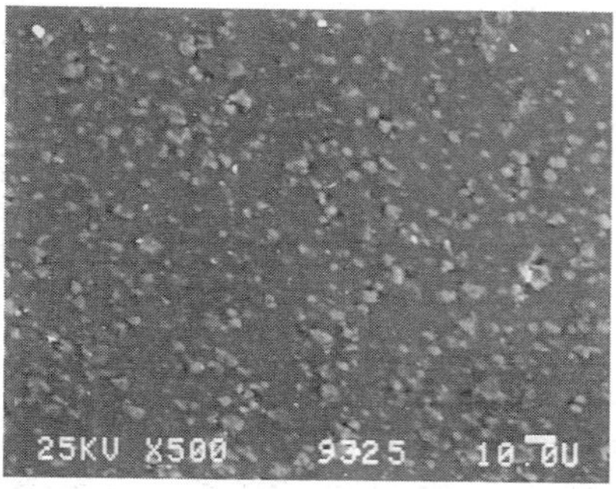

Surface of "B" Unexposed

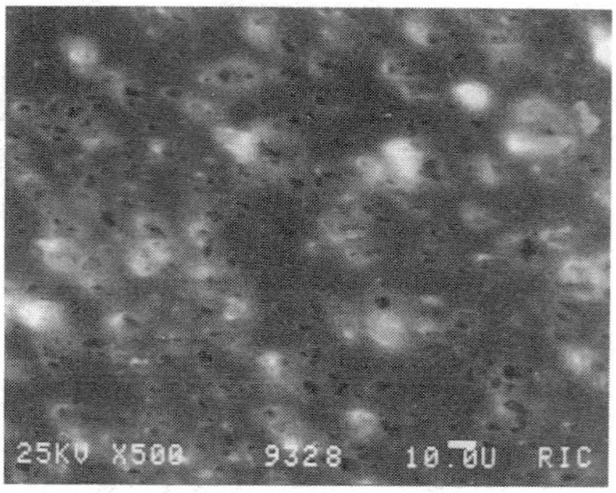

"B" Exposed 14 Days

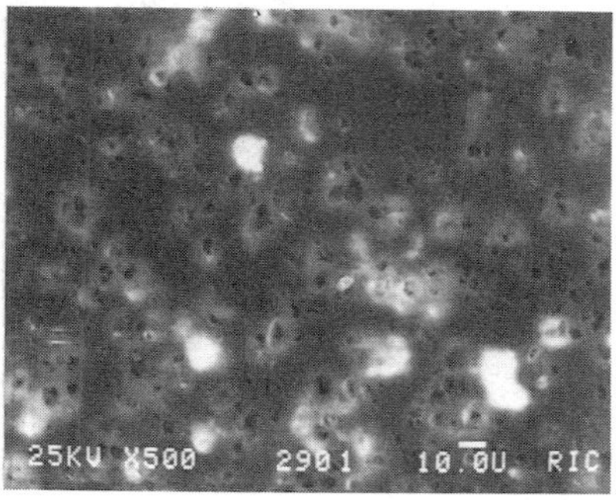

"B" Exposed 30 Days

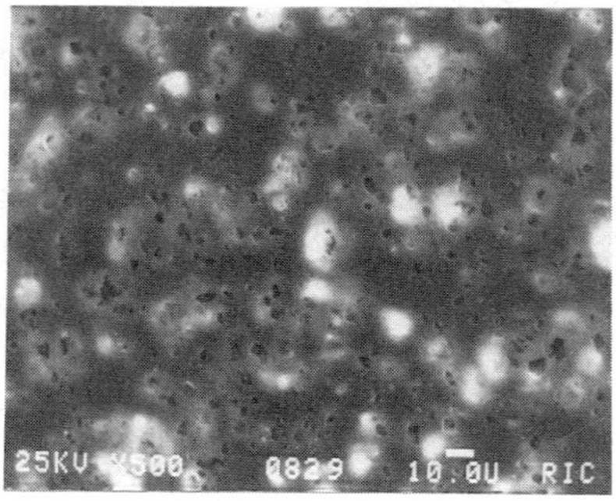

"B" Exposed 90 Days

FIG. 6--500X SEM's of Surface of Formulation "B" Before and After 14, 30 and 90 Days of H_2SO_4 Exposure

The photomicrographs in Figure 6 for 30 and 90-day exposure samples show the same amount of $CaCO_3$ attack as is observed in the 14-day exposure sample. Based on these photomicrographs, all the $CaCO_3$ at the surface available for attack, has been removed by the H_2SO_4 within 14 days.

Figure 7 shows the photomicrographs of the cross section of the 60 phr filled UPVC after 14, 30 and 90-day exposure to H_2SO_4. Area 1 (0.11mm below the surface) was selected for Figure 7 because it is closest to the surface, making it the most susceptible to attack.

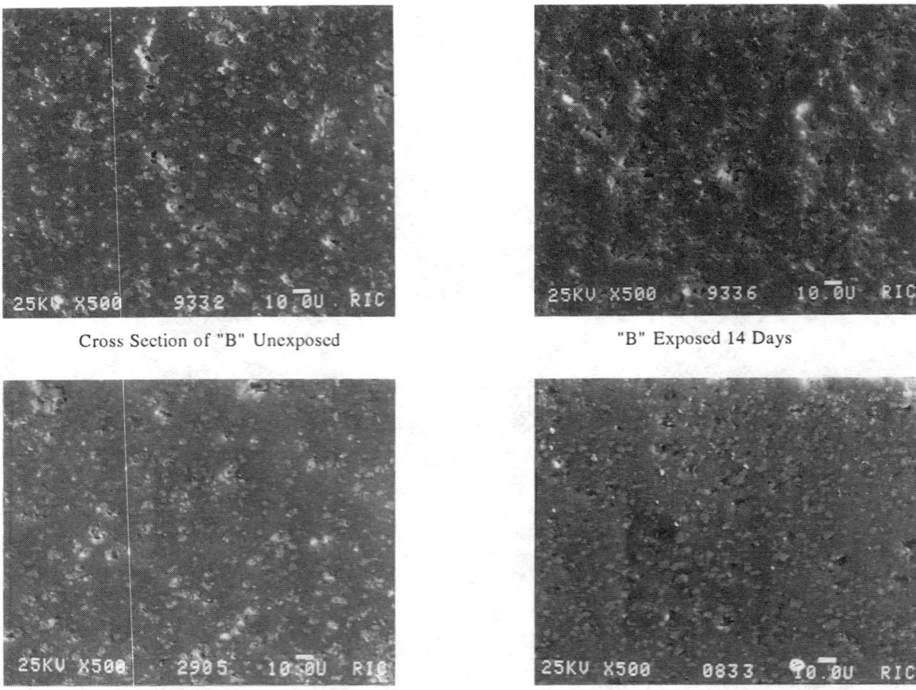

Cross Section of "B" Unexposed

"B" Exposed 14 Days

"B" Exposed 30 Days

"B" Exposed 90 Days

FIG. 7--500X SEM's of Cross Section Area 1 of Formulation"B" Before and After 14, 30 and 90 Days of H_2SO_4 Exposure

The photomicrographs in Figure 7 show no visible attack of the $CaCO_3$ just below the surface after up to 90 days of exposure to H_2SO_4. This further confirms the premise that the $CaCO_3$ is being protected from the H_2SO_4 by a barrier matrix of PVC polymer and that even after 90 days exposure, it has not penetrated that barrier.

Table 3 shows the WDS sulfur and calcium analysis data on the 60 phr filled specimens after 14, 30 and 90 days H_2SO_4 aging. Included in the table is data for both the surface and cross sectional areas discussed earlier in the paper.

The surface analysis data on extended H_2SO_4 exposure in Table 3 indicates there is no additional $CaCO_3$ attack after 14 days. This is evidenced by the fact that after the initial 14-day change in sulfur and calcium counts, essentially no additional changes are observed.

The WDS results on the sub-surface sections of the 60 phr filled UPVC show no indication, that with up to 90 days aging, the H_2SO_4 has not penetrated the PVC polymer matrix to attack the $CaCO_3$. Both the sulfur and calcium counts remain essentially unchanged, from 0 to 90 days exposure.

TABLE 3--Sulfur and Calcium WDS Analysis Results for 60 phr Sample Aged in 20 Weight Percent Sulfuric Acid

Area Analyzed	Distance Below Surface[1]	Aged	WDS Sulfur Counts			WDS Calcium Counts		
			14 Day	30 Day	90 Day	14 Day	30 Day	90 Day
Surface	---	No	6	---	---	5619	---	---
Surface	---	Yes	19	10	22	2739	2893	2452
Cross Section	0.11mm (0.004")	No	5	---	---	4713	---	---
Area 1	0.11mm (0.004")	Yes	9	4	7	4565	3624	4609
Area 2	0.03mm (0.012")	Yes	6	5	8	4932	4768	5195
Area 3	0.56mm (0.022")	Yes	4	4	7	4793	5105	5742
Area 4	0.83mm (0.033")	Yes	5	6	6	4776	5124	5692
Area 5	1.12mm (0.044")	Yes	4	5	4	4893	5012	5714

[1]Mid-Point of Cross Sectional Area

Both the SEM and WDS analysis show that only the $CaCO_3$ exposed on the surface of the 60 phr filled compound is attacked by the H_2SO_4 in 14 days or less. The $CaCO_3$ below the surface shows no indication of being attacked in up to 90 days aging in H_2SO_4. This confirms that the PVC matrix is not allowing the H_2SO_4 to penetrate below the surface and attack the $CaCO_3$.

CONCLUSION

The short and longer term aging effects of H_2SO_4 on two commercially used $CaCO_3$ filled UPVC sewer pipe compounds were studied. The scanning electron microscope was used to visibly observe the effects wavelength dispersive x-ray analysis was used to quantify the effects. Based on the results of the work, the following conclusions were drawn:

1. The visually observed effects of the H_2SO_4 on the $CaCO_3$ in the UPVC formulations correlated well with the calcium and sulfur quantitative analysis results.

2. The exposed $CaCO_3$ at the surface of the UPVC did react with the H_2SO_4 leaving voids.

3. $CaCO_3$ that is encapsulated and protected by the matrix of PVC polymer is not affected by the H_2SO_4. This includes the protected $CaCO_3$ on the surface as well as all the $CaCO_3$ contained in the body of the sample.

4. The effects of the H_2SO_4 on the $CaCO_3$ filler were observed to be the same in both the 25 and 60 phr filled UPVC compounds.

5. The study demonstrates both visually and analytically that the concerns for possible sewer gas acids attacking the $CaCO_3$ in UPVC sewer pipes are unfounded.

ACKNOWLEDGEMENTS

The authors wish to thank Mr. Eric Tomko for the help he provided in the compounding, molding and special sample preparations required for this study. Acknowledgement is also given here to Mr. Bill Butler of Ricerca, Inc. for his contributions in performing all of the scanning electron microscopy analysis as well as the wavelength dispersive x-ray spectroscopy microanalysis work. Finally, the authors wish to thank the Lamson Vylon Pipe division of The Lamson & Sessions Co. for their permission to publish this work.

REFERENCES

[1] ASTM D1785-91 "Standard Specification for Poly (Vinyl Chloride) (PVC) Plastic Pipe, Schedules 40, 80 and 120", 1993 Annual Book of ASTM Standards, Vol. 08.04, pp. 68-79

[2] ASTM D1784-92 "Standard Specification for Rigid Poly (Vinyl Chloride) (PVC) Compounds and Chlorinated Poly (Vinyl Chloride) (CPVC) Compounds", 1993 Annual Book of ASTM Standards, Vol. 08.04, pp. 64-67

[3] ASTM D3734-89 " Standard Specification for Type PSM Poly (Vinyl Chloride) (PVC) Sewer Pipe and Fittings", 1993 Annual Book of ASTM Standards, Vol. 08.04, pp. 392-400

[4] ASTM D2729-89 "Standard Specification for Poly (Vinyl Chloride) (PVC) Sewer Pipe Fittings", 1993 Annual Book of ASTM Standards, Vol. 08.04, pp. 286-289

[5] ASTM F679-89 "Standard Specification for Poly (Vinyl Chloride) (PVC) Large-Diameter Plastics Gravity Sewer Pipe and Fittings", 1993 Annual Book of ASTM Standards, Vol. 08.04, pp. 882-886

[6] ASTM F789-89 "Standard Specification for Type PS-46 Poly (Vinyl Chloride) (PVC) Plastic Gravity Flow Sewer Pipe and Fittings", 1993 Annual Book of ASTM Standards, Vol. 08.04, pp. 927-934

[7] ASTM F794-91 "Standard Specification for Poly (Vinyl Chloride) (PVC) Ribbed Gravity Sewer Pipe and Fittings Based on Controlled Inside Diameter", 1993 Annual Book of ASTM Standards, Vol. 08.04 pp. 935-940

[8] Schafer, P.L., Horner, I.S. and Wetzgall, R.A., "Case History of Sulfide Corrosion", Pipeline Design and Installation, American Society of Civil Engineers, March, 1990, pp. 168-193

[9] Bishop, R. R., "Retention of Pipe Stiffness for Polyvinyl Chloride (PVC) Pipe Samples Exposed to Various Environments and Constant Strain", Buried Plastic Pipe Technology, ASTM, October, 1990, pp. 7-20

[10] PVC Plastic Pipe, Section 20-17, Standard Specifications for Public Works Construction, BNI, 1991, pp. 234-237

Jun Tohda[1], Liming Li[2], and Hiroshi Yoshimura[3]

ANALYSIS OF THE FACTORS IN EARTH PRESSURE AND DEFORMATION OF BURIED FLEXIBLE PIPES THROUGH CENTRIFUGE MODEL TESTS

REFERENCE: Tohda, J., Li, L., and Yoshimura, H., "**Analysis of the Factors in Earth Pressure and Deformation of Buried Flexible Pipes Through Centrifuge Model Tests,**" Buried Pipe Technology: 2nd Volume, ASTM STP 1222, David Eckstein, Ed., American Society for Testing and Materials, Philadelphia, 1994.

ABSTRACT: A series of centrifuge model tests was conducted to estimate parametrically effects of the following factors on earth pressure and deformation of buried flexible pipes: pipe flexibility, pipe installation type, ground condition, and three burial dimensions (cover height, ditch width and bedding thickness). The tests yielded accurate measured data of circumferential distributions of normal and tangential earth pressures on the model pipes, circumferential distributions of bending strains in the pipe wall, and vertical pipe deflection, fully quantifying the effects of the investigated factors. It was found that Japanese current design standards, based on Marston-Spangler theory, do not conform to the test results by ignoring the effects of almost all the factors on the earth pressures, and, as a result, they specify definitely unsafe design values of bending moment and pipe deflection for the case when the pipes are buried by the open excavation method using sheet-piling.

KEY WORDS: flexible pipe, earth pressure, deformation, centrifuge test, pipe flexibility, ground condition, installation type, burial dimensions

The behavior of earth pressure and deformation of buried pipes is complicatedly affected by various factors such as construction conditions and properties of both soil and pipe. Nevertheless, effects of these factors on it have not been sufficiently quantified yet, owing to both the lack of a basic concept to explain comprehensively their roles, and the difficulty in measuring accurately earth pressure on buried pipes, in particular, on flexible pipes. Hence, the Japanese current design standards, based on traditional ultimate-equilibrium methods such as Marston-Spangler theory [1], can not correctly define their effects on factors of safety of buried pipes, yielding serious accidents in Japan.

In Japan, pipes having large diameters are usually buried in ditches constructed by open excavation methods using sheet-piling. In the latter-half of 1970's, one of the authors experienced many accidents in such constructions of sewer concrete pipes that a number of the pipes cracked when sheet-piles were extracted after backfilling. He conducted a full scale-field test [2] and a series of centrifuge model tests [3] to

[1]Associate Professor, [2]Postgraduate Student, Civil Engineering Department, Osaka City University, Osaka, Japan.

[3]Research Engineer, Research Institute of Technology, Konoike Construction Co., Ltd.

investigate the cause of the accidents, confirming the followings: 1) the sheet-pile extraction induces significant earth pressure concentration on the pipe top and bottom, together with an increase in total vertical earth pressure carried by the pipe, to cause the cracks of the concrete pipes, 2) this pressure concentration phenomenon always exists in higher or lower intensity in any types of pipe installations, 3) the traditional earth pressure theories can not explain the mechanism of this phenomenon, because they deal with only the total load on the pipes by assuming uniformly distributed vertical earth pressures, and 4) the conventional design procedures, built in accordance with such theories, do not correspond to the actual behavior of buried pipes. Furthermore, a theoretical elastic analysis [4] and FE elastic analyses [5] showed clearly the mechanism and critical factors of the earth pressure concentration, yielding a new basic concept that the earth pressure problem of buried pipes should be treated by elastic theory, contrarily to the traditional theories. Thus, in the rigid pipe case, the role of factors was systematically defined by using this concept, and the earth pressure for design was classified on the basis of these over 10 years' experimental and analytical studies [6].

Following these studies, the authors began a study on earth pressure and deformation of buried flexible pipes, because a serious accident took place in construction of flexible pipes in Japan, which was also caused by the sheet-pile extraction, in the similar way in the rigid pipe case. In the new study, they made efforts first for developing a technique to measure accurately earth pressure acting on flexible pipes, because reliable data had not been obtained in the flexible pipe case owing to the difficulty in measurement due to the pipe deformation, being the major obstacle in this field. This target has been accomplished through centrifuge model tests, carried out in 1990 [7].

This paper reports a series of centrifuge models tests using this measuring technique, conducted to investigate parametrically effects of the following factors on the behavior of buried flexible pipes: pipe flexibility, pipe installation type, ground condition, and three burial dimensions (cover height, bedding thickness and ditch width). In these tests, circumferential distribution of normal and tangential earth pressures on model pipes, circumferential distribution of bending strains in the pipe wall, and vertical pipe deflection were measured very accurately. The test results quantified the effects of the investigated factors sufficiently, indicating the necessity to revise the inadequate Japanese current design standards for buried flexible pipes. Since the test procedure has been detailed in the literature [7], it will be briefed in this paper.

OUTLINE OF CENTRIFUGE MODEL TESTS

Fig. 1 shows three aluminum model pipes, whose specifications are shown in Table 1. They are named as Flexible-pipe, Medium-pipe and Rigid-pipe, according to their flexibilities. The surfaces of these pipes are finished to be smooth, because this surface condition has been confirmed to be close to that of actual pipes, made of metal, plastic and concrete. The external diameter of the pipe, D, is 9 cm for Flexible-pipe and Rigid-pipe, and 8.6 cm for Medium-pipe. Their middle portions are divided into either 40 or 20 segments, each of which is connected by means of a small prismatic bar to a supporting beam fixed on the pipe-parts as shown in the figure. Strain gauges are glued on both sides of the prismatic bars to measure their axial and bending strains, through which earth pressures acting normally and tangentially on the pipe surface, σ and τ, are obtained respectively. Strain gauges are also glued on the inner and outer surfaces of the pipe-part to measure bending strains produced in the pipe wall. A deflection gauge is mounted to

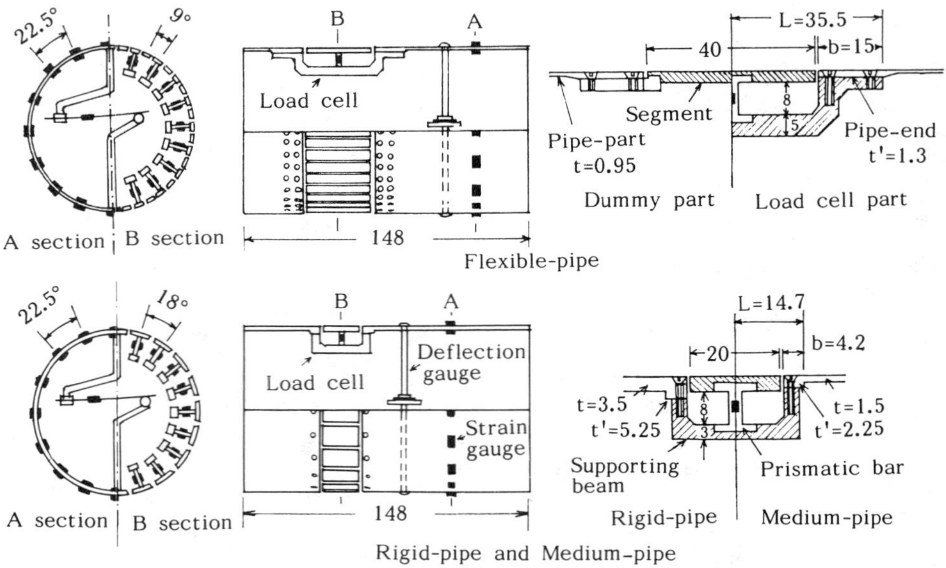

FIG. 1--Model pipes and detail of load cells (unit: mm).

TABLE 1--Dimensions of Model Pipes.

Pipe	External diameter D cm	Wall thickness t cm	Young's modulus E_p kgf/cm² (GPa)	Poisson's ratio ν_p	Flexural stiffness[1] S_f/a^3 kgf·cm²/cm (kPa/cm)	Weight of pipe gf/cm (N/cm)
Flexible-pipe	9	0.095			0.65 (64)	31 (0.30)
Medium-pipe	8.6	0.15	740 000 (72.6)	0.33	2.94 (288)	20 (0.20)
Rigid-pipe	9	0.35			32.6 (3 190)	35 (0.34)

[1] $S_f=E_pT^3/12/(1-\nu_p^2)$: a=D/2. Similarity law is (S_f/a^3) in models=(S_f/a^3) in prototypes.

measure the vertical deflection of the pipe.

Fig. 2 shows 2-D models that simulate the following three types of pipe installations: ditch-type with sheet-piling (Ditch-S, the most common installation type in Japan), embankment-type (Embk.), and ditch-type without sheet-piling (Ditch-0). These models are constructed to be 1/30 of the test prototypes using Rigid-pipe and Flexible-pipe,and 1/31.4 of those using Medium-pipe. The table in Fig. 2 shows the dimensions of prototypes corresponding to standard models having a cover height H=D, a thickness of soil bedding H_b=4/9·D, a ditch width B_d=13/9·D, and a width of the ground B=34/9·D. In the Ditch-S model, a pair of plain aluminum plates with a smooth surface is installed in a homogeneous model ground as model sheet-piles; their thicknesses were determined to be 15 cm in prototype scale considering the thickness of soil adhering to the concave parts of the extracted sheet-piles, which is usually observed in the actual construction. At both ends of the model ground in the Ditch-S and Embk. models, a pair of rigid steel plates with a smooth surface is installed as boundaries. The ditch in the Ditch-0 model is formed by a

Dimensions of prototype corresponding to standard models (m)

D	H	H_b	B_d	B
2.7	2.7	1.2	3.9	10.2

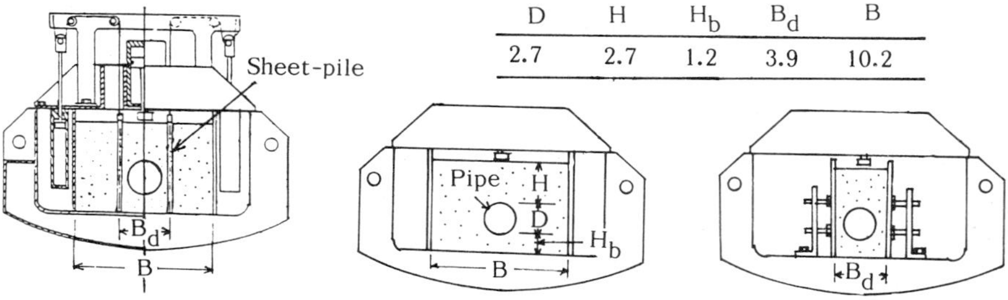

FIG. 2--Models for the three types of pipe installations.

pair of rigid steel plates with a rough surface, assuming that the natural ground does not deform during construction.

The models were put into a centrifugal acceleration field of 30 G or 31.4 G (G: gravitational acceleration) by using a centrifuge shown in Fig. 3. In the Embk. and Ditch-0 models, a load equivalent to the design load of a 20-ton vehicle was applied at the center of the ground surface by means of a strip loading plate 2 cm wide. In the Ditch-S models, a pair of the model sheet-piles was extracted simultaneously, and the surface load was applied afterwards. The simultaneous sheet-pile extraction method was adopted both to produce high earth pressure concentration on Rigid-pipe similar to that observed in the former field test [2], and to simplify the behavior of the model pipes.

Three kinds of soils: dry sand, decomposed granite and silty sand, whose properties are shown in Table 2, were used as ground materials. The main difference among them is the content of the fine graded fraction with grain size under 0.075 mm. The decomposed granite is usually used as the backfill material in pipeline construction in Japan. Dense and loose grounds were prepared by pouring for the tests using the dry sand; only loose ground was prepared by light compaction every 2 cm layers, for the tests using both the decomposed granite with a water content w=10 % and the silty sand with w=12 %. The pouring and compaction of the soils was performed in the direction of the longitudinal axes of the model pipes. The properties of the model grounds are shown in Table 3, together with the respective friction parameters against the pipe surface.

Table 4 shows the conditions and contents of the test series Ⓐ, Ⓑ and Ⓒ. The investigated factors, combined with the types of pipe installations, are: pipe flexibility and density of the dry sand in series Ⓐ, ground material in series Ⓑ and three burial dimensions (H, H_b and B_d) in series Ⓒ. Since the standard models are common in all the test series, total number of the tests is counted as 36.

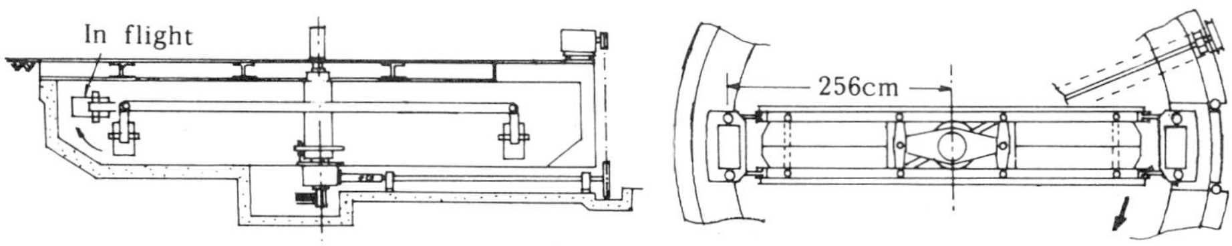

FIG. 3--Centrifuge.

TABLE 2--Properties of ground materials.

Soil (Classification)	G_s	Grain size distribution Maximum mm	>75μm %	75μm> %	5μm> %	U_c	ρ_{dmax} g/cm³	ρ_{dmin} g/cm³	w_{opt} %
Dry sand (SP)	2.65	1.4	100	0	0	2	1.58	1.32	-
Decomposed granite (SF)	2.71	2.0	83.7	8.6	7.7	70	1.92	1.37	11.4
Silty sand (SF)	2.67	2.0	69.6	16.9	13.5	115	1.86	1.18	13.5

TABLE 3--Properties of model grounds.

Soil	Density	Water content w %	ρ_d g/cm³	D_c[1] %	D_r %	Shear strength[2] c_d kgf/cm² (kPa)	ϕ_d degree	Friction[3] c_p kgf/cm² (kPa)	ϕ_p degree
Dry sand	Dense	0	1.55	97	83	0 (0)	43	0 (0)	17
	Loose	0	1.43	91	47	0 (0)	37	0 (0)	16
Decomposed granite	Loose	10	1.50	78	30	0.09 (8.8)	38	0.01(1.0)	24
Silty sand	Loose	12	1.50	81	58	0.30 (29.4)	32	0 (0)	25

[1] $D_c=\rho_d/\rho_{dmax}$, [2] strength under consolidated-drained condition obtained through direct shear test, and [3] friction parameters against the smooth pipe surface obtained through direct shear apparatus.

TABLE 4--Conditions and contents of the tests.

Test series	Flexibility of pipe	Ground condition Soil	Density	Type of pipe installation	Cover height H (cm)	Bedding thickness H_b (cm)	Ditch width B_d (cm)	Number of tests
Ⓐ	Flexible Medium Rigid	Dry sand	Dense Loose	Ditch-S Embk. Ditch-0	9	4	13	18
Ⓑ	Flexible	Dry sand Decomposed-granite Silty sand	Loose	Ditch-S Embk.	9	4	13	6
Ⓒ-1	Flexible	Dry sand	Loose	Ditch-S Embk. Ditch-0	4.5 9 18	4	13	9
Ⓒ Ⓒ-2	Flexible	Dry sand	Loose	Ditch-S Embk.	9	1 2 4	13	6
Ⓒ-3	Flexible	Dry sand	Loose	Ditch-S Ditch-0	9	4	13 17 21	6

TEST RESULTS FOR THE DITCH-S STANDARD MODELS IN SERIES Ⓐ

Since change in data due to the sheet-pile extraction is rather complicated, it is represented first by using measured results for the standard Ditch-S models in series Ⓐ.

Fig. 4 shows vertical deflections ΔD of the three model pipes, measured for the dense and loose dry-sand grounds. The symbols, ⓐ to ⓕ, in the sheet-pile extraction process denote the test stages when the lower ends of the sheet-piles passed through the levels ⓐ to ⓕ shown in the explanatory figure. The greater the pipe flexibility is, the greater ΔD it generates. After the test stage ⓔ, the ΔD values do not increase in any of the cases, even when the surface load is applied.

Fig. 5 illustrates distributions of σ and τ on Rigid-pipe and Flexible-pipe in polar coordinates, measured at five test stages in the sheet-pile extraction process, being compressive σ and downward τ counted as positive. The left-half and right-half of each figure shows the data for the dense and loose ground, respectively. The pressure σ on Rigid-pipe concentrates remarkably on the pipe top and bottom owing to the sheet-pile extraction; the magnitude of this earth pressure concentration is close to that observed in the former field test [2]. On the other hand, the maximum σ on the upper-half and lower-half of Flexible-pipe

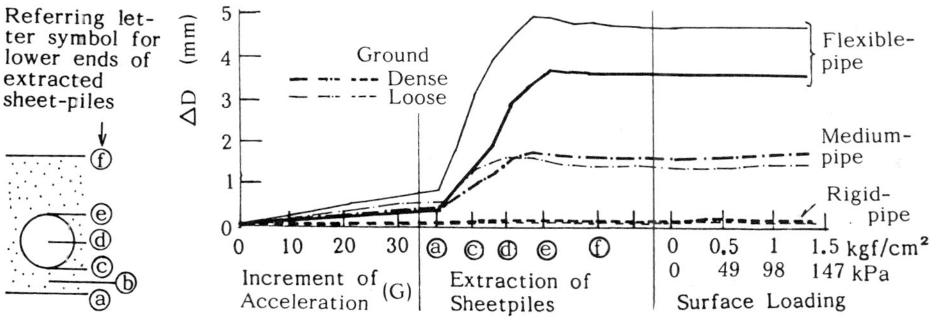

FIG. 4--Change in vertical pipe deflection during the Ditch-S standard tests.

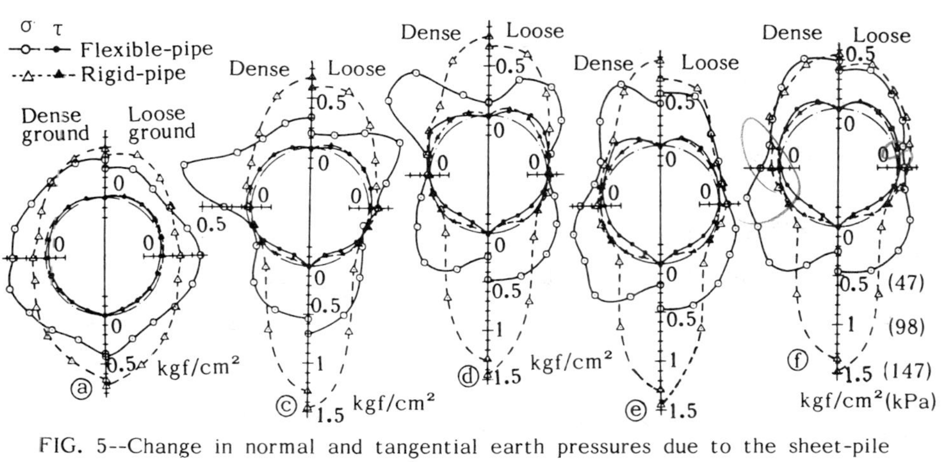

FIG. 5--Change in normal and tangential earth pressures due to the sheet-pile extraction, measured in the Ditch-S standard models.

change their acting positions upwards with the sheet-pile extraction, resulting in smaller σ at the pipe top and bottom than at adjacent areas. Although the pressures τ are considerably smaller than σ in any cases, they increase with the sheet-pile extraction, more remarkably in Flexible-pipe than in Rigid-pipe. Medium-pipe generated an intermediate changing pattern of σ and τ between the two patterns in the Rigid-pipe and Flexible-pipe cases (the data are omitted here).

The marks in Fig. 6 show bending strains for 9 points on the three model pipes, measured at the test stage ⓔ; the horizontal axis denotes the angle measured from the pipe top. The greater the pipe flexibility is, the greater bending strains it generates.

High accuracy in the measurements of the earth pressures, bending strains and pipe deflections was confirmed as follows:

1. The force equilibrium (ΣV=ΣH=ΣM=0) was proved to be fully satisfied, when the measured earth pressures as shown in Fig. 5 and the pipe weights were applied as external forces.

2. Distributions of bending strains were calculated through the principle of minimum work under the same external load conditions. The lines illustrated in Fig. 6 are examples of distributions of the calculated bending strains in the Ditch-S models, showing an excellent agreement with the measured ones, plotted as the marks.

3. Pipe deflections were calculated from distributions of the measured bending strains through the principle of virtual work. The calculated values were almost coincident with the measured ones.

4. Similar confirmations were made in all other tests.

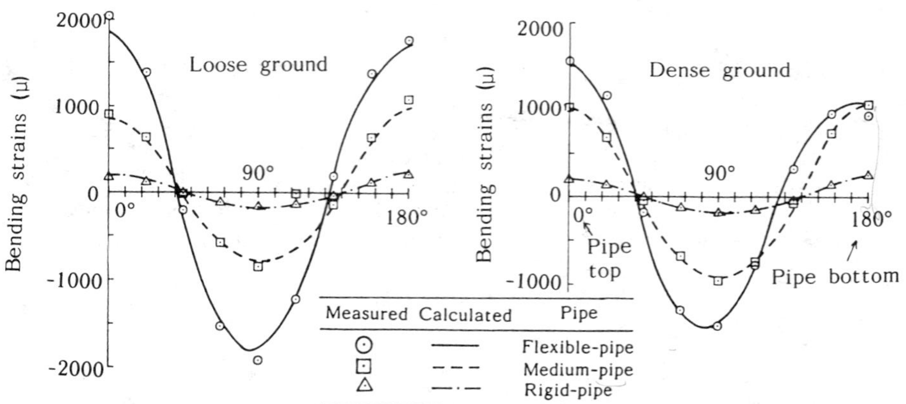

FIG. 6--Measured and calculated bending strains in the Ditch-S standard models.

MEASURED NORMAL AND TANGENTIAL EARTH PRESSURES

Fig. 7 to Fig. 11 illustrate the measured distributions of σ and τ in the whole test series in polar coordinates; the data in the Ditch-S models, except in Fig. 8, are those measured at the test stage ⓔ during the sheet-pile extraction. These figures show the following effects of the investigated factors as:

Pipe Flexibility and Density of The Dry Sand (Fig. 7)--In every types of pipe installations, the greater pipe flexibility generates the more uniform distribution shape of σ, except for the only an unique case of Flexible-pipe in the Ditch-S model. The difference in density of the

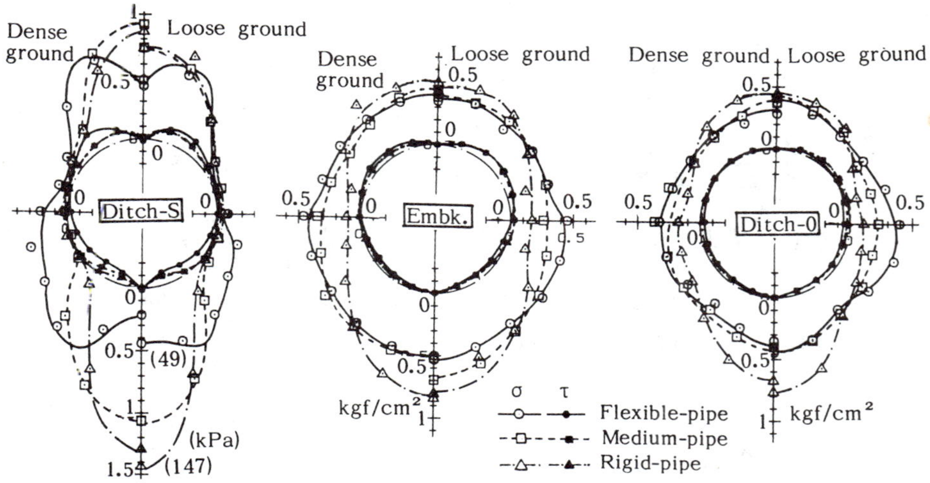

FIG. 7--Distributions of measured earth pressures for different pipe flexibilities in series Ⓐ.

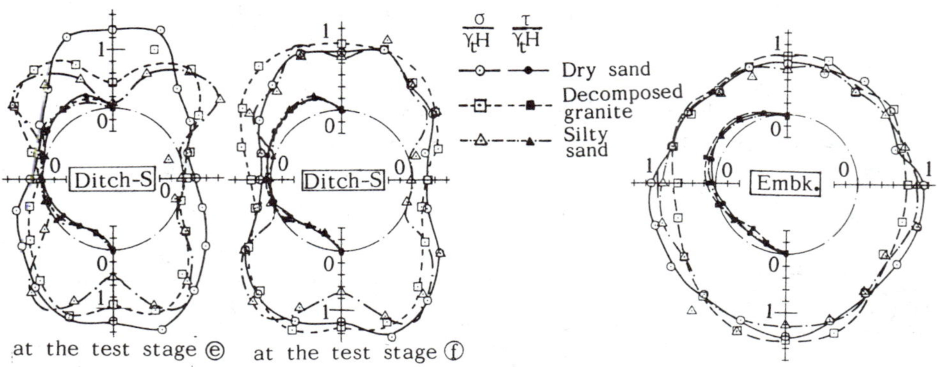

FIG. 8--Distributions of measured earth pressures for different ground materials in series Ⓑ.

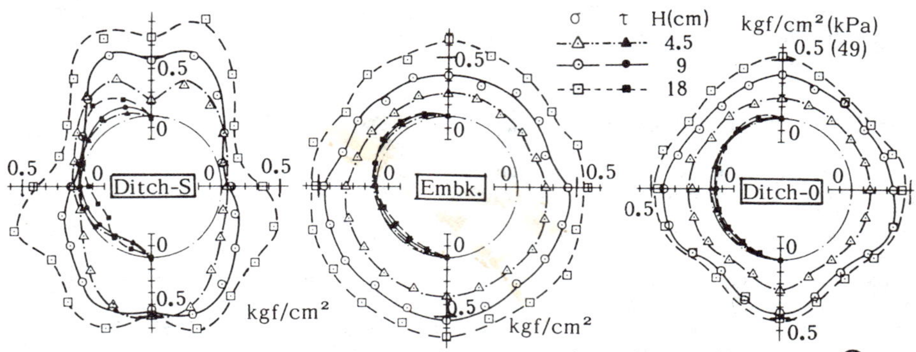

FIG. 9--Distributions of measured earth pressures for different H in series Ⓒ-1.

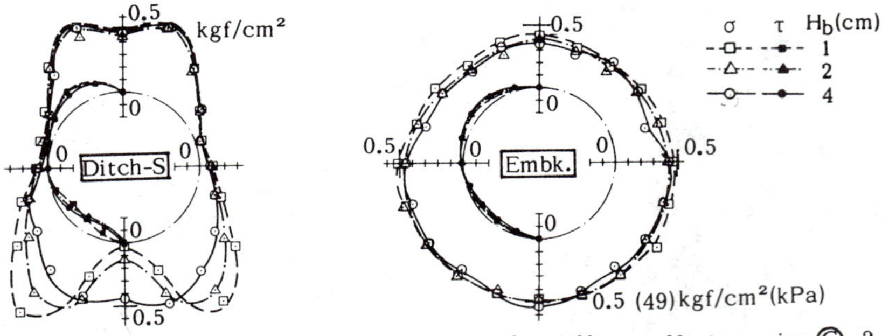

FIG. 10--Distributions of measured earth pressures for different H_b in series ©-2.

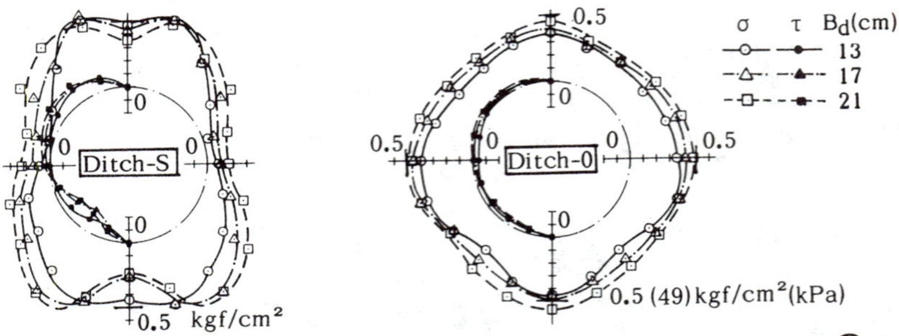

FIG. 11--Distributions of measured earth pressures for different B_d in series ©-3.

dry sand slightly affects the earth pressure distributions.

Ground Material (Fig. 8)--σ and τ in the figure are normalized by the overburden pressures $\gamma_t H$ (γ_t: wet unit weight of the soils, $\gamma_t=\gamma_d$ in the dry sand ground, H: cover height), to eliminate the difference in γ_t of the three ground materials. In the Ditch-S models, the different ground materials generate quite different distributions of σ at the test stages ⓔ and ⓕ. In the Embk. models, the decomposed granite generates somewhat long and slender distribution shape of σ, while the other two soils generate more uniform distribution shapes of σ.

Burial Dimensions (Fig. 9 to Fig. 11)--The earth pressures depend on the H, H_b and B_d values to great extents in any models, except for the the data in the Embk. model with different H_b in the test series ©-2.

DISTRIBUTIONS OF MEASURED VERTICAL AND HORIZONTAL LOADS

The measured σ and τ are translated into vertical load p_v, vertical reaction load p_r, and horizontal load p_h through the following equations:

$$p_v \text{ and } p_r = \sigma + \tau \cdot \tan\theta, \quad p_h = \sigma - \tau \cdot \cot\theta$$

in which θ is the angle measured from the pipe top. The left-half of each figure in Fig. 12 to Fig. 16 illustrates these measured loads, normalized by the overburden pressure γH ($\gamma=\gamma_t$ in the test series B and $\gamma=\gamma_d$ in the other test series using the dry sand), showing the effects of the investigated factors as follows:

<u>Pipe Flexibility and Density of The Dry Sand</u> (Left-half of each figure in Fig. 12)--The data shown in Fig. 12 are for the loose dry sand ground; the dense ground generated similar tendencies as described here. In the Ditch-S model, Flexible-pipe generates an unique M-letter-shaped distribution of p_v and p_r, while Rigid-pipe and Medium-pipe generate mountain-shaped distributions as a result of high pressure concentration on the pipe top and bottom; distribution shapes of p_h are concave in any pipes. In the Embk. and Ditch-0 models, the greater pipe flexibility generates the more uniform load distribution shape.

<u>Ground Material</u> (Left-half of each figure in Fig. 13)--The different ground materials generate distribution shapes of p_v, p_h and p_r

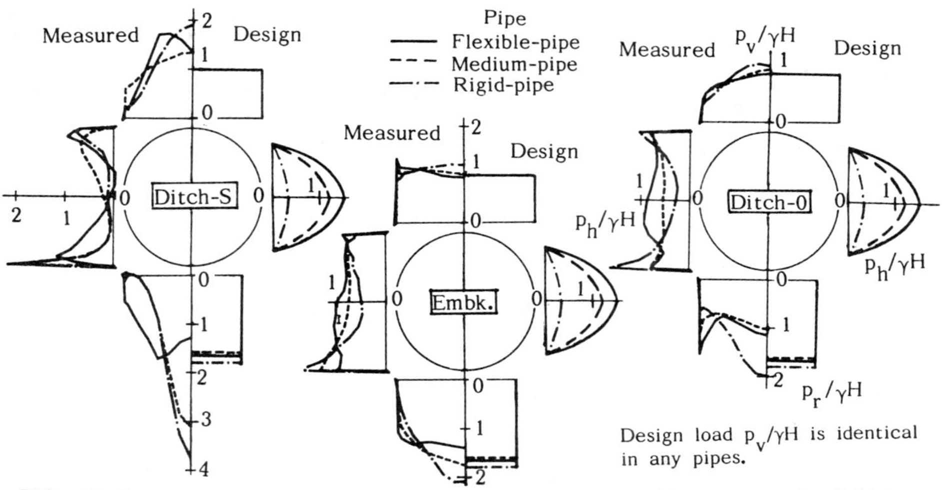

FIG. 12--Distributions of measured and design loads for different pipe flexibilities in series Ⓐ (Loose ground).

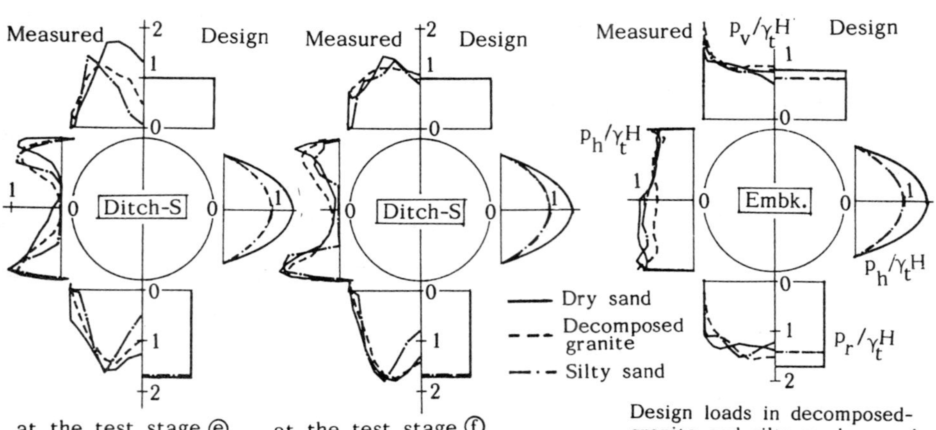

FIG.13--Distributions of measured and design loads for different ground materials in series Ⓑ.

Design loads in decomposed-granite and silty-sand grounds are identical. Design load $p_v/\gamma_t H$ in the Ditch-S model is identical in any soil grounds.

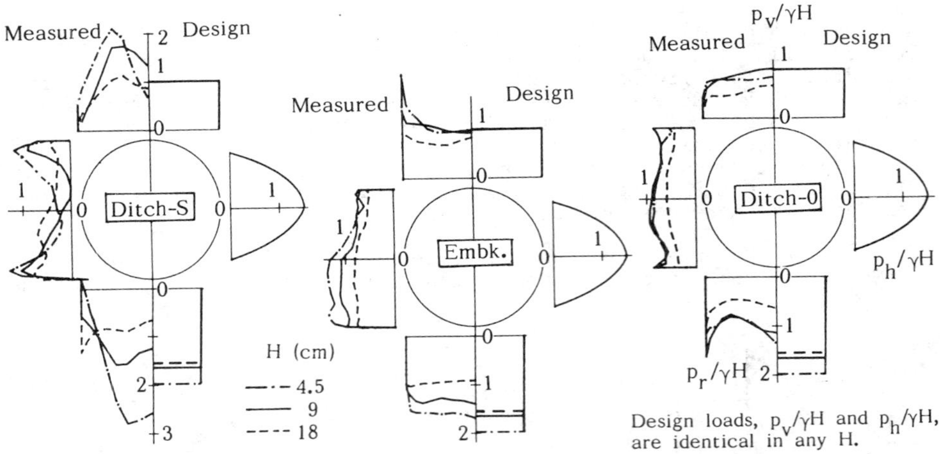

FIG. 14--Distributions of measured and design loads for different H in series © -1.

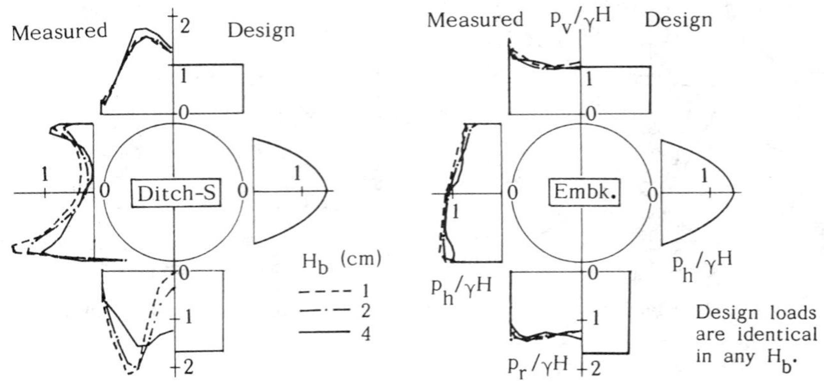

FIG. 15--Distributions of measured and design loads for different H_b in series © -2.

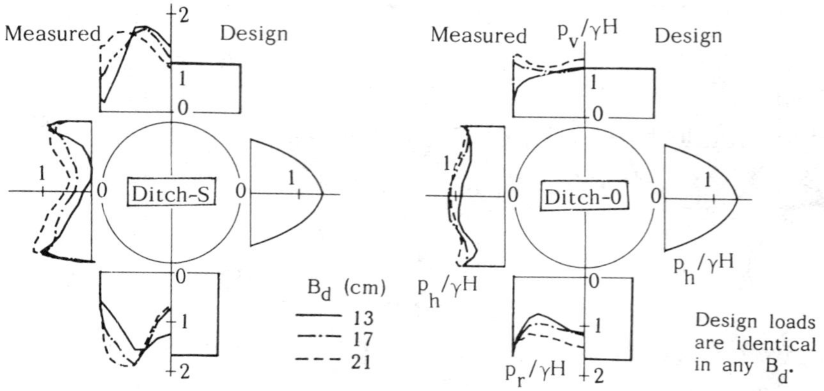

FIG. 16--Distributions of measured and design loads for different B_d in series © -3.

having different positions of their peak values in the Ditch-S model, in particular, at the test stage ⓔ. In the Embk. model, the different ground materials generate different load distribution shapes of p_h and p_r only on the lower-half of the pipe.

Burial Dimensions (Left-half of each figure in Fig. 14 to Fig. 16)--With an increase in H, the load distribution shapes in the Ditch-S models tend to be more uniform, while only the load intensities decrease everywhere in the Embk. and Ditch-0 models (cf. Fig. 14). H_b affects the load distribution shapes only on the lower-half of the pipe in the Ditch-S model (cf. Fig. 15). Different B_d create different load distribution shapes in both Ditch-S and Ditch-0 models (cf. Fig. 16).

MEASURED TOTAL LOAD, BENDING MOMENT AND PIPE DEFLECTION

Marks in the top, second and lowest rows of Fig. 17, except for the marks x on the thick lines, illustrate nondimensionally the following measured results: i) total vertical and horizontal loads $P_v/\gamma HD$ and $P_h/\gamma HD$ obtained by integrating the measured loads, ii) bending moment at the pipe top and bottom K ($=M/\gamma HR^2$, M: bending moment obtained from the measured bending strains, R: neutral radius of the pipe), and iii) vertical pipe deflection $\Delta D/2R$, where $\gamma=\gamma_t$ in the test series Ⓑ. The data in the Ditch-S models correspond to those measured when the maximum pipe deflection was recorded in the sheet-pile extraction process: at the test stage ⓔ for the dry sand grounds, and at the test stage ⓕ for both decomposed-granite and silty-sand grounds. Ignoring several exceptional data, the measured results show the followings:

The Total Load $P/\gamma HD$--The $P_v/\gamma HD$ values are greater than 1 in the Ditch-S and Embk. models, and smaller than 1 in the Ditch-0 models. The $P_h/\gamma HD$ values are considerably smaller in the Ditch-S models, and slightly smaller in the other two models, than the $P_v/\gamma HD$ values.

The Bending Moment K and The Pipe Deflection $\Delta D/2R$--The Ditch-S models always generate considerably greater K and $\Delta D/2R$ values than the other two models. Furthermore, the changing ranges of K and $\Delta D/2R$, caused by the difference in each investigated factor, are always greater in the Ditch-S models than in the other two models.

The measured results in the test series Ⓐ also show that the density of the dry sand affects the K and $\Delta D/2R$ values in the Ditch-S model using Flexible-pipe (cf. the most left-side data in series Ⓐ).

COMPARISON BETWEEN TEST RESULTS AND DESIGN

Here, let us compare the test results with design values obtained by the design standard of MAFF [8]. This MAFF standard is one of the most representative standards in Japan, and its core is built on the basis of Marston-Spangler theory. It specifies that:

1. The load p_v distributes uniformly over the width of the pipe. Its intensity is calculated through $p_v=\gamma_t H$ in the Ditch-S type, Marston's projection formula in the Embk. type, and the smaller one between $p_v=\gamma_t H$ and Yansen's formula in the Ditch-0 type. When applying the test conditions, the intensity of p_v became to be identical with $\gamma_t H$ in any cases, except in the two cases of the decomposed-granite and silty-sand grounds in the Embk. model.

2. The load p_r distributes uniformly over the width of bedding; its intensity is calculated as the sum of P_v and pipe weights. The bedding

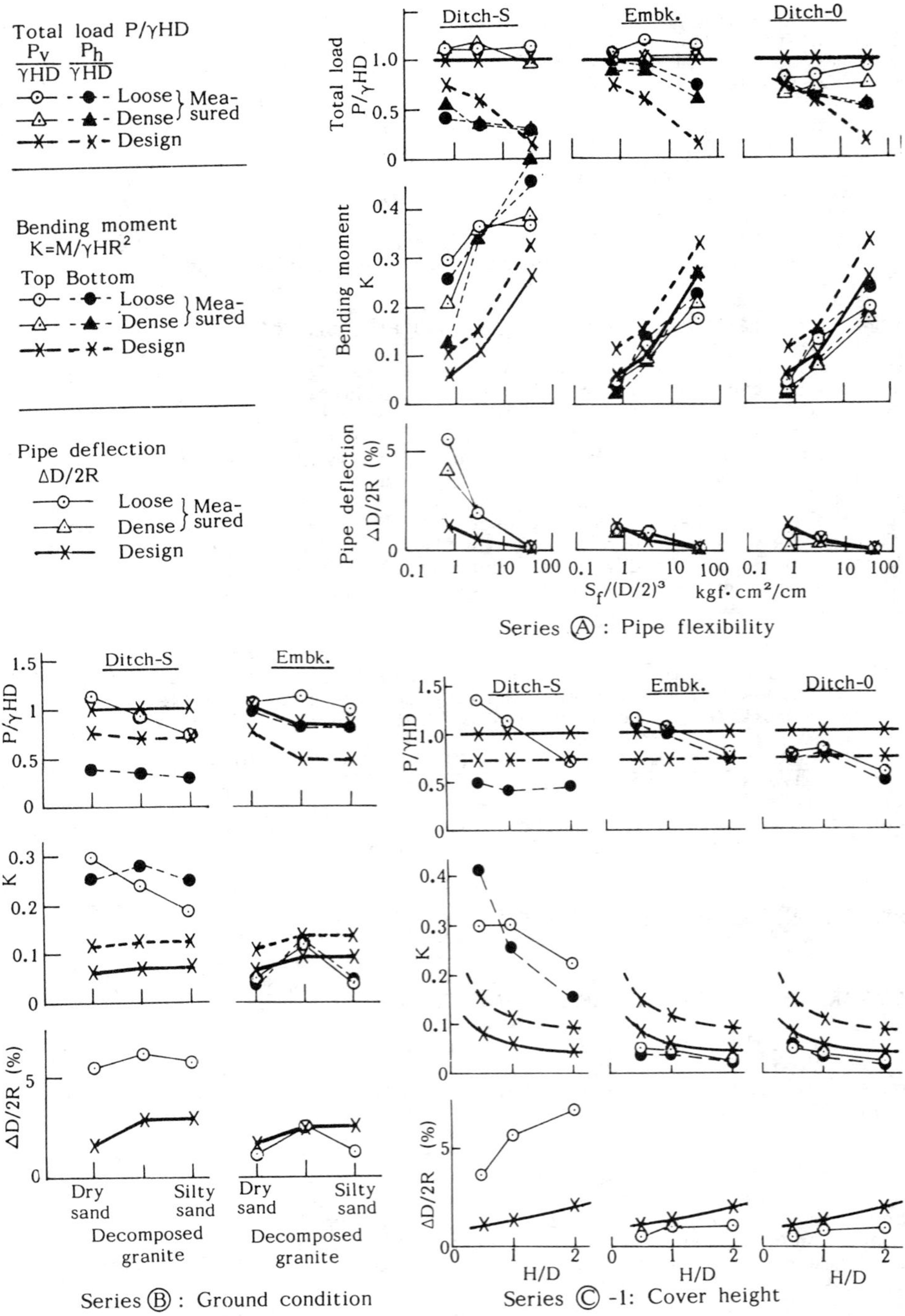

FIG. 17--Comparison between the test results and design.

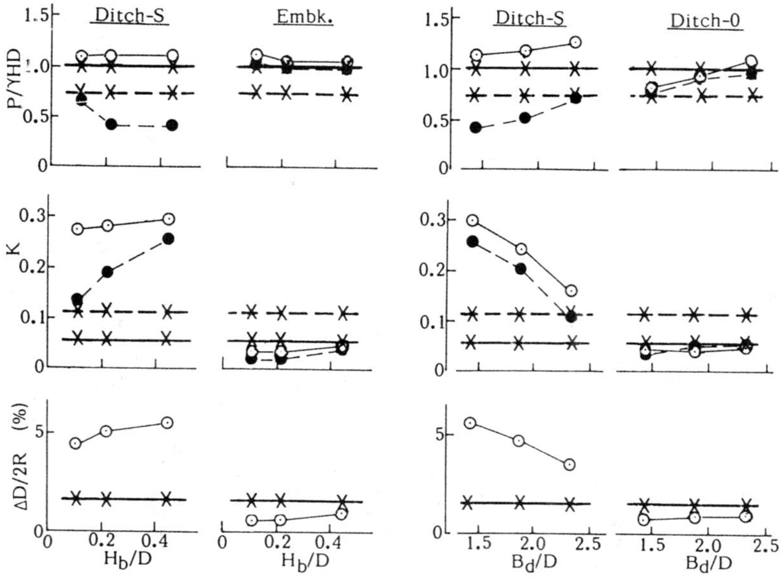

Series Ⓒ -2: Bedding thickness   Series Ⓒ -3: Ditch width

FIG. 17--Comparison between the test results and design (continue).

angle was specified as 90 degrees for all the present ground conditions.

3. The load p_h distributes parabolically over the middle 100 degrees. Its maximum intensity is calculated through $e'/R \cdot \Delta x/2$, where e': modulus of passive earth pressure, Δx: horizontal pipe deflection due to only p_v, and R: neutral radius of the pipe. The value of e' is specified as 140 kgf/cm² (13.7 MPa) for the dense and loose dry-sand ground conditions, and e'=28 kgf/cm² (2.74 MPa) for both decomposed-granite and silty-sand ground conditions.

The right-half of each figure in Fig. 12 to Fig. 16 illustrates the non-dimensional design loads, calculated in accordance with the MAFF standard under the test conditions. Comparison between the measured and design loads shows that: i) the design loads, being almost identical in any test series, definitely ignore the effects of all the investigated factors on the measured load distributions, ii) distribution shapes of the design loads do not conform to the measured ones in any cases, and iii) the parabolic shapes of the horizontal design loads p_h are definitely erroneous in the Ditch-S types to compromise the safety of the pipes.

Design values of P/γHD, K and ΔD/2R are respectively plotted as the marks, x, on the thick lines in the top, second and lowest rows of Fig. 17. Comparison between the design values and the measured ones shows that: i) the design values of $P_v/\gamma HD$ and $P_h/\gamma HD$ are unchanged in any cases, except in the cases of the test series Ⓑ, ignoring the effects of the investigated factors on the measured total loads, and ii) the design values of K and ΔD/2R are extremely smaller in the Ditch-S types and somewhat greater in the Embk. and Ditch-0 types, than the measured ones.

Thus, it is confirmed that the design standard not only specifies the behavior of buried flexible pipes utterly different from that obtained in the tests, but also it generates definitely unsafe design values for the Ditch-S type, so that it may cause the pipe damages in actual construction. The other Japanese design standards were confirmed to generate similar results to the ones described here.

CONCLUSIONS

A series of centrifuge model tests yielded accurate measured results to quantify successfully the effects of the following factors on earth pressure distribution, bending moment and pipe deflection of buried flexible pipes: pipe flexibility, type of pipe installation, ground condition, and three burial dimensions. The test results confirmed that flexible pipes deflect drastically at the sheet-pile extraction when the pipes are buried by open excavation method using sheet-piling (Ditch-S type of pipe installation). It was found that Japanese current design standards based on Marston-Spangler theory do not conform to the test results by ignoring the effects of almost all the investigated factors on the test results, and as a result, they specify definitely unsafe design values for the Ditch-S type to cause damages of flexible pipes in actual construction. In order to avoid the pipe damages due to the sheet-pile extraction, therefore, the current design standards should be revised on the basis of the test results represented in this paper as immediately as possible, and until this revision is performed, the countermeasures reported in the literature [9] should be employed in the actual design and construction works.

REFERENCES

[1] Spangler, M. G., "The Structural Design of Flexible Pipe Culverts," Bull.153, Iowa Engineering Experiment Station, Iowa State University, 1941.

[2] Tohda, J. et al., "Earth Pressure on Underground Concrete Pipe in A Field Test," Proceedings of ASCE International Conference on Advances in Underground Pipeline Engineering, 1985.

[3] Tohda, J., "Earth Pressure Acting on Buried Pipes," Developments in Geotechnical Aspects of Embankments, Excavations and Buried Structures, Balkema, Rotterdam, 1991.

[4] Tohda, J. et al., "A Study of Earth Pressure on Underground Pipes Based on Theory of Elasticity", Proceedings of JSCE, Vol.376, 1986.

[5] Tohda, J., Mikasa, M., and Hachiya, M., "Earth Pressure on Underground Rigid Pipes: Centrifuge Model Tests and FEM Analysis", Centrifuge 88, Balkema, Rotterdam, 1988.

[6] Tohda, J. et al., "FE Elastic Analysis of Measured Earth Pressure on Buried Rigid Pipes in Centrifuged Models," Proceedings of ASCE International Conference on Pipeline Design and Installation, 1990.

[7] Tohda, J. et al., "Earth Pressure Acting on Buried Flexible Pipes in Centrifuged Models," Proceedings of ASCE International Conference on Pipeline Design and Installation, 1990.

[8] Ministry of Agriculture, Forestry and Fisheries in Japan, Design Standard of Pipelines, 1988.

[9] Tohda, J., Yoshimura, H. et al., "Centrifuge Model Tests on Several Problems of Buried Pipes", Centrifuge 91, Balkema, Rotterdam, 1991.

Timothy J McGrath[1], Ernest T. Selig[2] and Leonard C. DiFrancesco[3]

STIFFNESS OF HDPE PIPE IN RING BENDING

REFERENCE: McGrath, Timothy J., Selig, Ernest T., and DiFrancesco, Leonard C., "**Stiffness of HDPE Pipe in Ring Bending,**" Buried Plastic Pipe Technology: 2nd Volume, ASTM STP 1222, Dave Eckstein,Ed., American Society for Testing and Materials, Philadelphia, 1994.

ABSTRACT: A ring bending test was devised to evaluate the behavior of plastic pipes that have been held under constant deformation and then subjected to intermittent loads. The apparatus consists of a frame capable of locking the test pipe at a fixed level of deflection in ring bending. A load cell monitors the change in force with passing time required to hold the deflection level. At preselected time intervals an additional deflection increment is briefly applied to the pipe and then released. The load required to produce this deflection increment is recorded. The pipe is then returned to the original level of deformation. Test results show that the long term decrease in load, or stress relaxation, is unaffected by the short term load increments. The short term modulus is much greater than the long term relaxation modulus and does not decrease with time. Test results also show that short term load increments may be treated as a new load on the pipe, with a stiffness response governed by the short term modulus.

KEY WORDS: long-term modulus, parallel plate, plastic pipe,relaxation, short-term modulus, viscoelastic, Young's modulus

High density polyethylene plastic (HDPE), from which buried pipe are made, is a viscoelastic material. As such, the material stiffness, as characterized by Young's modulus, is time-dependent. Under constant load or constant deformation the modulus appears to decrease with increasing time; however, this is an equivalent elastic modulus resulting from forcing the material to fit a linear elastic model.

There has been some disagreement among engineers on the appropriate value of elastic modulus to use in calculations that involve both short and long-term loading. One design situation where the appropriate modulus must be known is when buried pipe, under long-term

[1]Senior Associate, Simpson, Gumpertz and Heger, Inc., Arlington, MA 02174, and Doctoral Student, University of Massachusetts, Amherst, MA 01003.

[2]Professor of Civil Engineering, University of Massachusetts, Amherst, MA 01003.

[3]Graduate Student, Department of Civil Engineering, University of Massachusetts, Amherst, MA 01003.

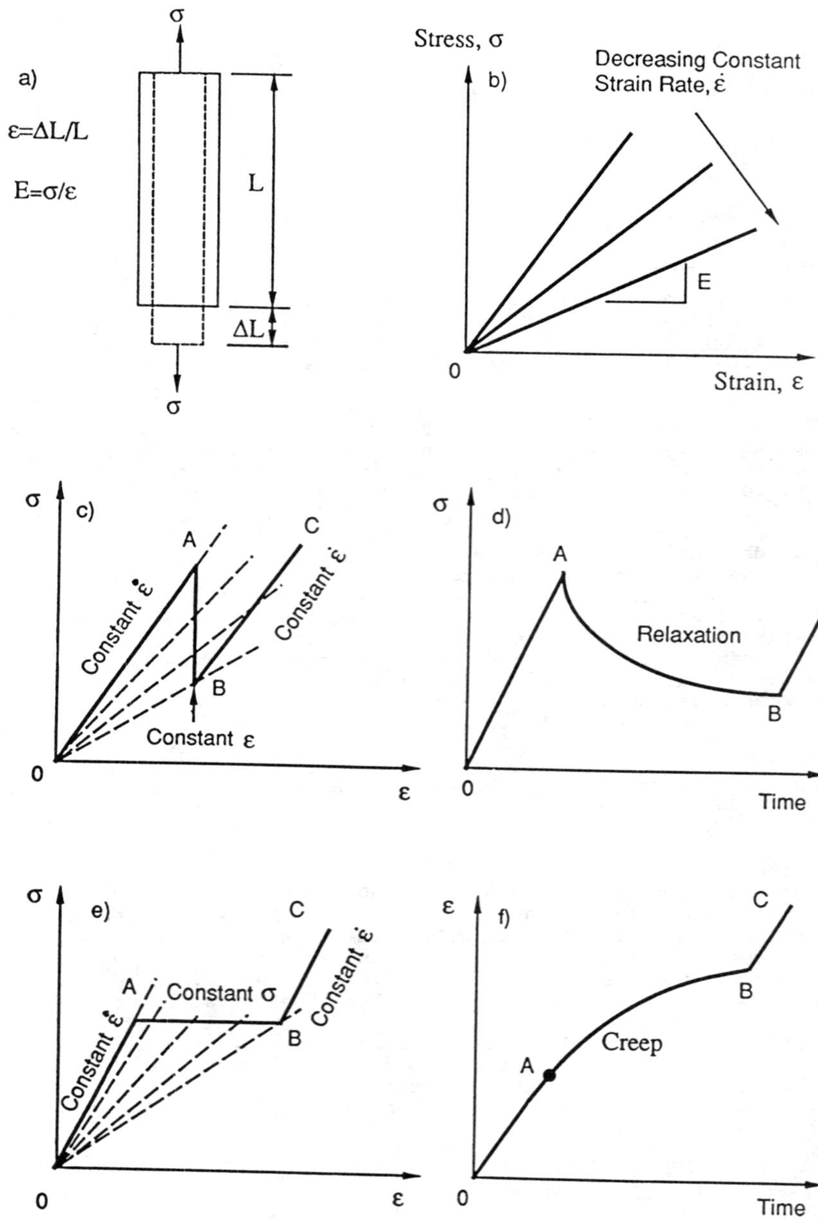

Fig. 1 Idealized Stress-Strain Characteristics of Linear Viscoelastic Behavior

earth load, are subjected to short-termvehicle loads. While the principle of superposition has been shown to be applicable to the behavior of thermoplastics in the linear viscoelastic range [1], plastic pipe are often strained beyond that range, and the validity of superposition is not assured (linear viscoelasticity means that the effect of any change in stress or strain is independent of stress or strain history and can be added algebraically to the effects of prior stress or strain increments). The validity of superposition is also called into question by the complex nature of the stresses in a deflecting ring. The stresses and strains are highest at the crown, invert and springline where the bending moment is greatest, but are quite low at the intermediate points where the bending moment is zero. Thus, during deflection, portions of the ring may pass into the non-linear viscoelastic range while other portion remain in the linear range.

To gain some insights into the nature of the HDPE modulus for corrugated pipe, parallel plate ring bending tests were conducted which combine constant deflection with periodic superimposed deflection cycles. This provides information on the change in equivalent elastic modulus with time for comparison with the equivalent modulus for superimposed short term load cycles. This paper describes the test procedure and gives sample results.

VISCOELASTIC BEHAVIOR

The effect of linear viscoelastic material behavior on the equivalent linear elastic Young's modulus is illustrated in Fig. 1. A uniaxial tension test is assumed (Fig. 1a), but the trend is the same in compression. At a constant rate of strain the Young's modulus will be nearly constant at low strain levels, but the value will decrease with decreasing strain rate (Fig. 1b). If the stress is increased at a constant strain rate, followed by a period when the strain is held constant, and then the stress is again increased at the same constant strain rate, the result is as shown in Figs. 1c and 1d. During the period of constant strain the stress is relaxing and the equivalent elastic modulus is decreasing. During the periods of constant strain rate the modulus is approximately constant and the same value in both cases (OA and BC) if the strain level remains in the linear viscoelastic range. For HDPE resins the linear limit is about 2% strain [1]. Thus the apparent material drop in modulus during the relaxation period does not affect the stress strain behavior when the loading is resumed. If, during the period between constant strain rate loadings, the material is held at constant stress, then the strain will creep as shown in Figs. 1e and 1f. The effect on the trend of equivalent elastic modulus will be the same as for the relaxation case.

The limit of the linear viscoelastic range in a parallel plate test is suggested by Hashash [2] as shown in Fig. 2, which presents the change in apparent Young's modulus with time for 24 in. diameter corrugated HDPE pipe loaded at the same stain rate and held at 5%, 10% and 15% deflection levels. The resin from which these pipes where manufactured is not known but they are believed to be similar to the resin used in the tests reported below. The figure shows that the modulus for the test at 5% deflection is notably higher than the modulus for the tests at 10% and 15% deflection. This indicates that between 5% and 10% deflection the linear viscoelastic range is exceeded at the crown, springlines, and/or invert. This causes a reduction in the modulus for tests that continue to higher displacements, but, interestingly there is no continuing loss of stiffness as the tests at 10% and 15% deflection indicate the same modulus at any given time.

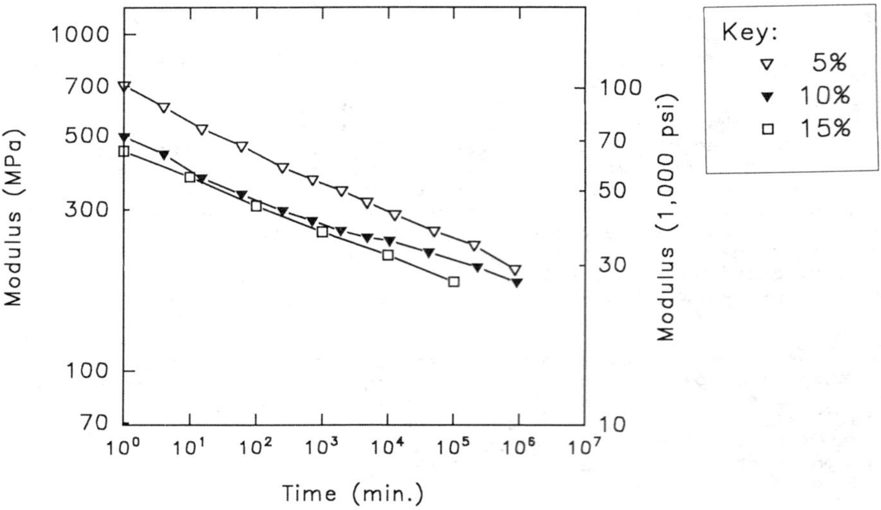

Fig. 2 Modulus Versus Time for Pipe Held at Different Deflection Levels [2]

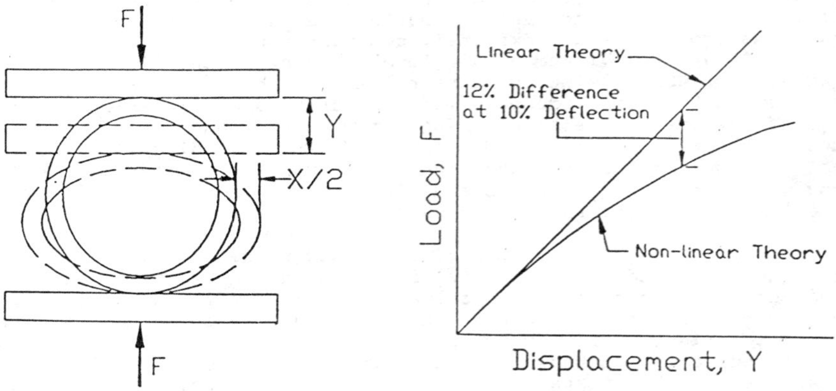

Fig. 3 Ring Bending Test

RING BENDING TEST THEORY

The ring bending test described in this paper is a variation of ASTM Standard Test Method for Determination of External Loading Characteristics of Plastic Pipe by Parallel Plate Loading (D2412). It is illustrated in Fig. 3. For a material with constant Young's modulus the load increases at a decreasing rate with increasing vertical displacement. This is a result of geometry change, i.e., the ring changes from a circular shape to an oval shape as deflection increases. Small deformation theory neglects the geometry change and hence gives a linear load deformation plot. The non-linear theory is appropriate for the tests described in this paper. For deflection levels considered in this paper the load versus horizontal deformation plot is linear.

The relationship between load and pipe deflection is as given by DIN Standard for Thermoplastic Pipe and Fittings with Profiled Outer and Smooth Inner Surfaces (DIN 16961) and [2]:

$$F = \frac{EI}{0.1488\ R^3}\ Y(C_1)\ , \qquad (1)$$

and

$$F = \frac{EI}{0.1366\ R^3}\ X\ , \qquad (2)$$

where

$$C_1 = \left(1 + \frac{0.2\ Y_m}{0.1488}\right)^{-1}\ , \qquad (3)$$

$$Y_m = Y/(2R)\ , \qquad (4)$$

F = applied load (N/mm),

E = pipe material modulus of elasticity (MPa),

I = wall moment of inertia per unit length of pipe (mm^4/mm),

R = mean pipe radius or distance from center of pipe to the centroidal axis of pipe wall (mm),

Y = change in vertical pipe diameter (mm),

X = change in horizontal pipe diameter (mm),

C_1 = parameter to consider geometry non-linearity.

When the pipe material is not linear elastic, as for HDPE, a secant Young's modulus, E, may be determined using Eqs. 1 and 2. This approach was used for the load-relaxation test representing the change in pipe deflection from an undeformed state to a particular constant deflection level.

A modulus representing the periodic short-term deflection cycles superimposed on the load-relaxation tests can be approximated by using the slope of the applied load versus vertical deflection curve at the midpoint of the increment. The slope of the curve defined by Eq. 1 is a function of the Young's modulus of the pipe material; however, when the actual, non-linear, geometry relationship is considered the slope of the curve changes, even for a constant modulus, and it is necessary to use the derivative of Eq. 1 to determine the tangent modulus for any given incremental load or deflection applied to the deformed pipe.

The derivative giving the slope of the load-deflection curve is

$$\frac{dF}{dY} = \frac{EI}{0.1488\ R^3}\left(C_1 - \frac{0.2\ Y_m C_1^2}{0.1488}\right), \qquad (5)$$

where dF/dY is the slope of Eq. 1 at the deflection level Y_m which is computed as the average deflection for the increment dY.

GEOMETRY OF TEST PIPE

The test pipes used as an example in this paper were made of high density polyethylene (HDPE), cell class 324420C as defined in ASTM D3350. It had an annular corrugation exterior with a smooth interior liner heat bonded to the corrugation. The nominal interior diameter was 610 mm (24 in.). The exterior diameter was 726 mm (28.6 in.). The pipe was cut on cross-sections passing through the valley of the corrugations where it is bonded to the liner, to provide a length of 610 mm (24 in.).

The calculated pipe wall properties are given in Table 1. Two categories are indicated. One assumes that the unbonded portion of the liner is effective in carrying load, and the other assumes that the

TABLE 1--Pipe wall section properties

Property		Liner Effective	Liner Ineffective	Percent Decrease
h	mm (in.)	58 (2.3)	58 (2.3)	--
y_m	mm (in.)	20 (0.80)	26 (1.01)	--
A	mm^2/mm ($in.^2/in.$)	8.3 (0.32)	6.3 (0.25)	24
I	mm^4/mm ($in.^4/in.$)	3100 (0.19)	2310 (0.14)	26
R	mm (in.)	325 (12.80)	331 (13.01)	--

h = total depth of section,
y_m = radial distance from inside wall to wall centroid,
A = wall cross-section area per unit length of pipe,
I = wall moment of inertia per unit length of pipe,
R = radius to centroid of pipe wall.

unbonded portion is not effective. Other tests, not reported in this paper, suggest that the unbonded liner does not contribute to resisting load, thus the properties under "liner ineffective" are used for all data reduction.

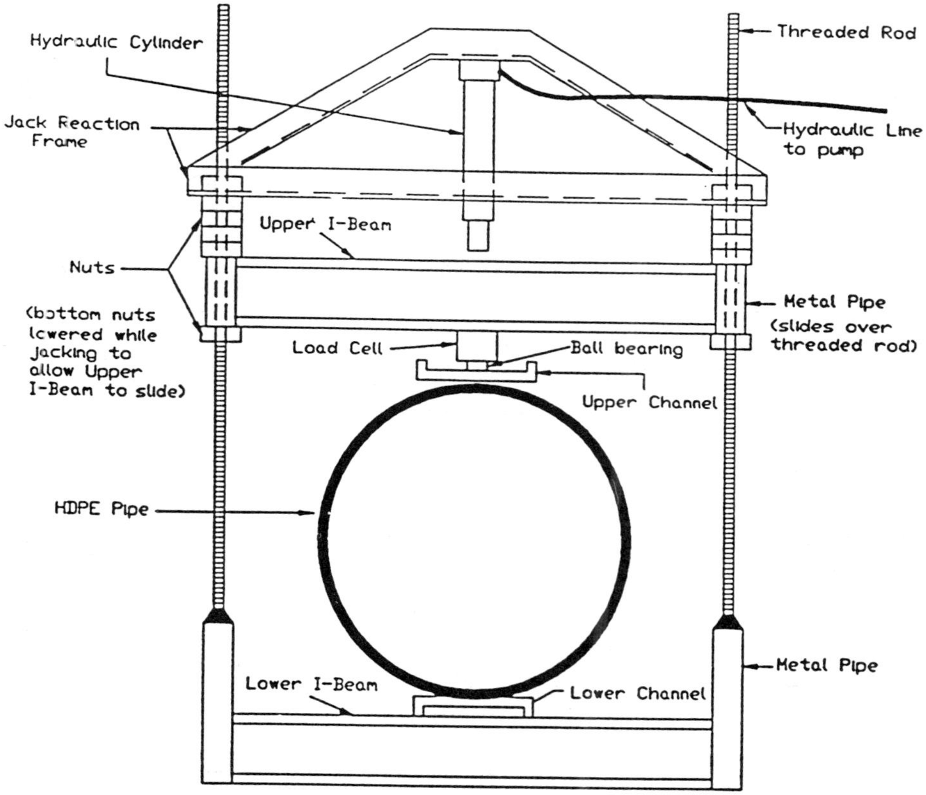

Fig. 4 Ring Bending Test Apparatus

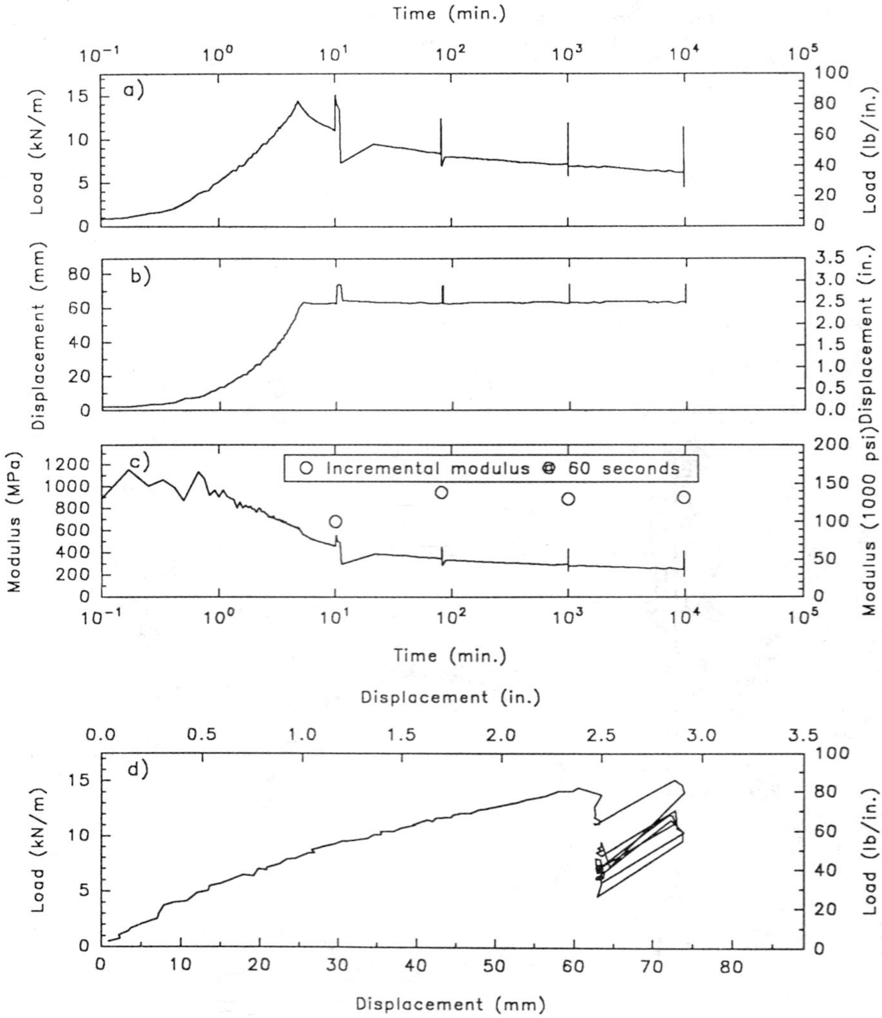

Fig. 5 Ring Bending Test Involving Initial Loading at 13 mm/min. Followed by Constant Deflection Load Relaxation with Superimposed Deflection Cycles

APPARATUS AND PROCEDURE

The apparatus for the ring bending test is shown in Fig. 4. Upper and lower I-beams, tied together with threaded rods, compress and clamp the pipe. Upper and lower channels distribute the clamping force along the length of the pipe. A removable hydraulic jack and reaction frame apply the force required to deflect the pipe. A load cell is used to measure the change in compression force required to maintain the pipe deflection state. Extensometers (not shown) are connected across the vertical and horizontal diameters to indicate the pipe deflections.

With the nuts under the upper I-beam released, the hydraulic jack is used to deflect the pipe to the prescribed deformation state. The lower and upper nuts are tightened against the upper I-beam to lock the deformation in place for the long term load relaxation tests.

In the example tests described in this paper a long-term constant vertical deflection equal to 10% of the inside diameter was used. This deflection level was imposed in most tests in approximately ten seconds, giving a deformation rate of about 360 mm/min (14 in./minute). The long term load was recorded starting at 60 seconds. Pipes were also deformed at the rate prescribed by ASTM D2412 of 13 mm/min (0.5 in./min) for comparison. In these tests a data acquisition system was used to continuously measure the load and deflection beginning at the start of the loading process.

The incremental deflection cycle tests are performed by lowering the bottom nuts and rapidly increasing the pipe deflection by a fixed amount. The load is monitored for 60 seconds and then the jack is released to return the pipe to its long term constant deflection state. The bottom nuts are again tightened.

This deflection cycle test can be superimposed on the long term relaxation test at any time. In the example test described in this paper the deflection increment was 10 mm (0.4 in.) which represents about 1.5% of the pipe original inside diameter. This deflection increment was applied in approximately seven seconds, and then load readings were recorded at thirty and sixty seconds after applying the incremental deflection. The incremental deflection was released after the sixty second reading was obtained, and the pipe immediately rebounded to its previous long-term deflection level.

EXAMPLE RESULTS

Figure 5 gives results obtained with the computer data acquisition system which clearly illustrate the pipe response to all stages of the test. Consider first the load vs. time and displacement vs. time plots (Figs. 5a and 5b). The initial portion of these curves represents loading at a deformation rate of 13 mm/min (0.5 in./min) until the 10% deflection is reached. This took about 5 minutes. Thereafter deflection was held constant and the load relaxed. At 10 minutes after the start of loading and at 3 later times a 10 mm deflection cycle was imposed. The increment was applied at the same rate in all cases. The corresponding loads were recorded. At the end of each deflection cycle the load returned to the original relaxation curve.

A plot of load against corresponding deflection from Figs. 5a and 5b is given in Fig. 5d. The secant slopes (F/Y) connecting the origin of the plot with successive points (F,Y) on the curve represent pipe stiffness change. These sets of points (F,Y) were used with Eq. 1 to calculate the corresponding equivalent secant Young's modulus values (E) for the pipe. As previously noted, the assumption was made that the unsupported portion of the liner was not effective in resisting the bending. The values of Young's modulus are indicated by the curve in Fig. 5c. As shown in Fig. 5c the secant modulus is around 1000 Ma for the first minute and then decreases continuously to about 600 Ma at the end of the loading stage (in 5 minutes). Thereafter the secant modulus gradually decreases over time during the relaxation stage. Small perturbations occur in the secant modulus during the displacement cycles when the total load and total displacement are considered.

The drop in modulus during loading is contrary to the theory of linear viscoelasticity which states that a test at a constant strain rate will have a constant modulus, as shown in Fig. 1. There are two reasons for this. First, as noted above, the state of stress in a ring under parallel plate deflection is non-uniform and the load-displacement curve is not linear (see Eq. 1). Thus a constant rate of deflection test does not produce a constant rate of strain in the pipe material. Second, based on Fig. 2, the linear viscoelastic limit is exceeded during the test and the lower modulus is thus a function of the higher stress level and so is lower in the more highly deflected specimen. The second reason is thought to be the more important one.

The change in load (ΔF) after 60 seconds divided by the 10 mm change in deflection (ΔY) for each cyclic deflection test is assumed to represent the slope dF/dY in Eq. 5. These slopes, then, are used in Eq. 5 to calculate the equivalent tangent Young's modulus (E) for the pipe material for each deflection cycle. The resulting values are plotted as circles on Fig. 5c. The tests at 100, 1,000 and 10,000 minutes all show that the pipe stiffness, as represented by the modulus in Fig. 5c is the same as that of the original pipe. This is the case even though during loading the linear viscoelastic limit was exceeded and the apparent

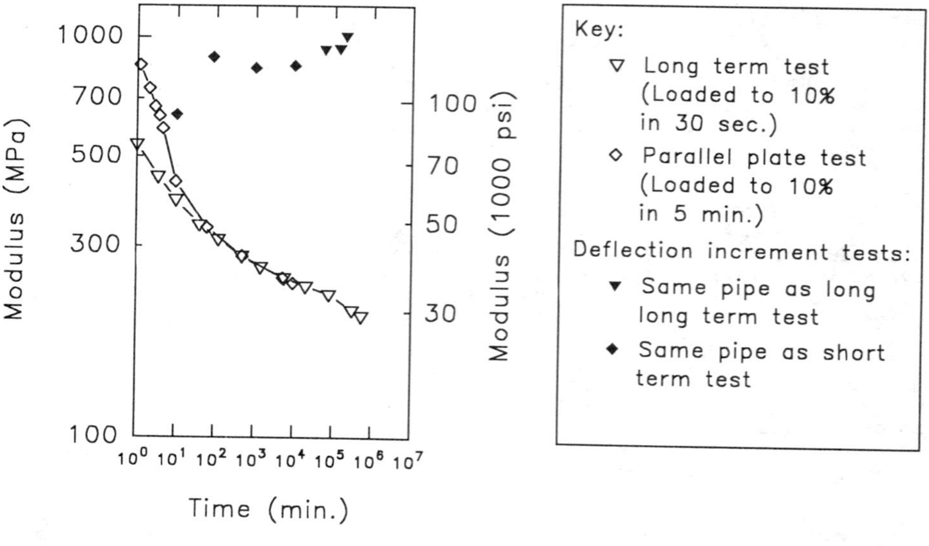

Fig. 6 Constant Ring Defleciton Tests (10%) Long and Short-Term Moduli

modulus reduced. The incremental short term modulus test at 10 minutes, however, indicated that the reduction of modulus due to exceeding the linear viscoelastic limit was still having some affect on the response of the pipe.

Most of the ring bending tests involved deflecting the pipe to 10% in 10 sec (360 mm/min) and beginning load relaxation measurements at 60 sec. The corresponding values of F and Y were used with Eq. 1 to get the secant Young's modulus. The values are given in Fig. 6 together with the data from Fig. 5c. Note that the lower loading rate gives a higher modulus initially, but the two curves converge at about 10 min. The reason for this is that the test at the slower load rate is at deflection level of less than 5%, the suggested linear viscoelastic limit from Fig. 2, for the first two and one half minutes of loading and during this time the modulus is represented by the higher curve in Fig. 2. After exceeding 5% deflection the behavior is represented by the lower curves in Fig. 2 and thus the two curves in Fig. 6 come together at about 100 min.

The short term modulus values from the superimposed deformation cycles are also shown on Fig. 6. These measurements were obtained over a period of more than 1 year from the start of the test. The short term modulus values were much higher than the secant modulus at the start of the relaxation test. Also the short term modulus values did not decrease over time with load relaxation.

SUMMARY AND CONCLUSIONS

Tests of HDPE plastic pipe in ring bending were conducted to evaluate the material modulus. In load relaxation tests at constant deformation the equivalent Young's modulus decreased continuously with time. However, superimposed deflection cycles gave much higher equivalent Young's modulus which did not decrease with load relaxation. This was true even when the prior deflection levels had exceeded the linear viscoelastic limit.

ACKNOWLEDGEMENTS

This research was initiated under the sponsorship of the Pennsylvania Department of Transportation and the U.S. Department of Transportation Federal Highway Administration. The research was continued under the sponsorship of the Corrugated Plastic Pipe Association. The equipment was fabricated in the University of Massachusetts Engineering Shops by technician Miles Eastman.

REFERENCES

[1] Heger, F.J., Chambers, R.E., and Dietz, A.G., "Structural Plastics Design Manual." ASCE Manual of Practice No. 63, ASCE, New York, 1984.

[2] Hashash, Naila, "Design and Analysis of Deeply Buried Polyethylene Drainage Pipes." Report No. PDT91-387D, Ph.D. Dissertation, Department of Civil Engineering, University of Massachusetts, September.

David T. Iseley,[1] **Mohammad Najafi**[2] **and Robert D. Bennett**[3]

TRENCHLESS PIPELINE REHABILITATION WITH PLASTIC MATERIALS

REFERENCE: Iseley, D. T., Najafi, M., and Bennett, R. D., **"Trenchless Pipeline Rehabilitation with Plastic Materials,"** Buried Plastic Pipe Technology: 2nd Volume, ASTM STP 1222, Dave Eckstein, Ed., American Society for Testing and Materials, Philadelphia, 1994.

ABSTRACT: This paper describes the state-of-the-art review for trenchless pipeline rehabilitation being conducted under the Construction Productivity Advancement Research (CPAR) program. For the purpose of this study, the existing trenchless rehabilitation methods will be classified into six groups: cured-in-placed pipe (CIPP), sliplining, in-line replacement, deformed & reshaped, point source repair, and sewer manhole rehabilitation. For each group, specific rehabilitation procedures will be provided and advantages and disadvantages of each method will be discussed. This information is being obtained through literature search and survey of manufacturers, utility owners and contractors of sewer rehabilitation products and services. However, the state-of-the-art review provides additional information such as an overview of design techniques with plastic pipes, the rehabilitation costs as a percentage of new installation and also to the amount of open-cut required for different methods. The intent of this state-of-the-art review is to provide information on capabilities of existing trenchless rehabilitation systems to assist design engineers and city officials to plan and conduct pipeline rehabilitation when dealing with a specific system.

KEY WORDS: pipeline rehabilitation, trenchless technology, pipe design, underground infrastructure, pipe installation, plastic pipe

---

INTRODUCTION

CPAR is a cost-shared research, development, and demonstration program authorized under section 7 of the Water Resources Development

---

[1]Associate Professor of Civil Engineering and Director of the Trenchless Technology Center (TTC), Louisiana Tech University, Ruston, LA 71272.

[2]Assistant Professor, Department of Engineering Technology, Missouri Western State College, Saint Joseph, MO 64507.

[3]Senior Research Civil Engineer, Geotechnical Laboratory, U.S. Army Corps of Engineers, Waterways Experiment Station, 3909 Halls Ferry Road, Vickburg, MS 39180.

Act of 1988. Under this act, U.S. Army Corps of Engineers (CORPS) has been authorized to undertake collaborative research and development with the U.S. construction industry to improve domestic construction productivity and cost effectiveness. This particular CPAR project is a two-year, $1.2 million project which began January 7, 1992. The primary partners for the project are the U.S. Army Engineer, Waterways Experiment Station and the Trenchless Technology Center (TTC) at Louisiana Tech University. The research partners of the project will evaluate trenchless methods, materials, and equipment, for installing and/or rehabilitating underground utility systems. Secondary industry partners (equipment and materials manufacturers, owners, contractors, and others) will be involved through the furnishing of materials, equipment, and manpower to conduct the necessary field and laboratory evaluations.

The state-of-the-art review is one of main elements of the research project, "Trenchless Construction: Evaluation of Methods and Materials to Install and Rehabilitate Underground Utilities." The research is a Corps and Industry cost-shared project funded under the Construction Productivity Advancement Research (CPAR) program. The laboratory partner is the U.S. Army Engineer Waterways Experiment Station (WES), Geotechnical Laboratory. The industry partner is the Louisiana Tech University's Trenchless Technology Center. Industry participants are contributing over half the total cost of this research. The overall objective of the CPAR program is to improve productivity in the U.S. construction industry, thereby enhancing the competitiveness of the U.S. industry in domestic and overseas markets. The Corps and other federal and state agencies will benefit by realizing reduced costs, enhanced safety and better products and services as the products of this research program are developed, tested, and commercialized.

This state-of-the-art review is being conducted to assist design officials and utility managers in selecting the most suitable trenchless rehabilitation alternatives for specific situations. The information in this review is for use by utility owners, city and state officials and government agencies, design engineers, manufacturers and suppliers, contractors and others who have some responsibility for planning, designing, constructing, financing, regulating, operating, or maintaining underground pipeline systems.

OBJECTIVES

The objectives of this review are as follows:

1. Identify some of the causes for pipeline deterioration and to determine advantages of trenchless pipeline rehabilitation.
2. To provide an inventory of available trenchless pipeline rehabilitation techniques in the United States.
3. To determine factors affecting selection of a specific technique, as well as performance criteria and installation requirements.
4. To present a basis for the design of trenchless pipeline rehabilitation systems using plastic materials.

TECHNICAL APPROACH

The technical approach for this research consists of:

1. Preliminary literature search and consultation with industry experts on the methods and materials for trenchless pipeline rehabilitation to provide a focus for this state-of-the-art review.
2. In-depth literature search and review process, focused on available techniques. Preparation of industry questionnaire.

3. Consultation with recognized pipeline rehabilitation experts and equipment manufacturers.

3. Survey of major contractors, manufacturers, and suppliers of pipeline rehabilitation materials.

4. Survey of professional, trade and research associations regarding existing research, existing and proposed regulations, and technical literature, including contract specifications.

5. Survey of city and local officials, utility owners, and regulatory agencies regarding their experience with trenchless pipeline rehabilitation.

6. Preparation of draft report.

7. Review of draft report by industry experts, coordinated by North American Society for Trenchless Technology (NASTT).

8. Preparation and publication of final report.

ADVANTAGES OF TRENCHLESS PIPELINE REHABILITATION

Trenchless pipeline rehabilitation with plastic pipe is rapidly growing in North America as demand creates an increasing need for cost-effective pipe rehabilitation. A recent study shows that the sewer rehabilitation market for the years 1993 to 1997 totals more than $3 billion dollars [1].

Many methods of trenchless pipeline rehabilitation are currently available or under development. Worldwide, there are presently over 50 different types of pipeline rehabilitation methods with over 500 different contractors providing services for these techniques [2]. These methods can be used to replace, upgrade, or renovate existing pipelines. They also can be used, within limits of technology, to enlarge existing pipes. This study provides an overview of these methods and presents guidance on the situations where each method is applicable.

Trenchless methods of pipeline rehabilitation with plastic pipe have many advantages over conventional excavation and replacement methods. Trenchless pipeline rehabilitation:

- Minimizes the need to disturb existing environment, traffic, or congested living and working areas.
- Uses predetermined paths provided by existing piping, reducing the steering and control problems associated with new routing.
- Requires less space underground, minimizing chances of interfering with existing utilities or abandoned pipes.
- Provides the opportunity to upsize a pipe (within the technology limits) without the problems associated with open-trench construction.
- Requires less exposed working area, therefore, is safer for both workers and the community.
- Minimizes the need for spoil removal, damage to pavements (the life expectancy of pavements was observed to have been reduced by up to 60% with dig-up repairs [3]), and disturbance to other utilities.
- By minimizing interference with existing utilities and environment, provides opportunities for greater construction productivity.

The basic trenchless pipeline rehabilitation methods can be categorized into the following types:

1. Cured-In-Place Pipe (CIPP)
2. Sliplining
3. In-Line Replacement
4. Deformed and Reshaped
5. Point Source Repair
6. Manhole Rehabilitation

Table 1 provides a summary of these basic methods, with additional information provided in the following sections. The various rehabilitation groupings are shown diagrammatically in Figure 1.

CURED-IN-PLACE PIPE (CIPP)

The CIPP method, invented in England by Insituform in 1977, is a thermoset resin-saturated material that is inserted into the existing pipeline by hydrostatic or air inversion or by mechanical pulling with a winch and cable. Recently a number of CIPP systems have been introduced into the market. This increased number of available methods provides many technical and economical benefits for utility owners through increased competition.

The primary components of the CIPP are a flexible fabric tube and a thermosetting resin system. For typical CIPP applications, the resin is the primary structural component of the system. These resins generally fall into one of the following generic groups, each of which has distinct chemical resistance and structural properties. All have good chemical resistance to domestic sewage:

1. Unsaturated Polyester
2. Vinylester
3. Epoxy

Unsaturated polyester resins were originally selected for the first CIPP installations due to their chemical resistance to municipal sewage, good physical properties in a CIPP composite, excellent working characteristics for CIPP installation procedures, and economic feasibility. Unsaturated polyester resins have remained the most widely used systems for the CIPP processes for over two decades.

Vinylester and epoxy resin systems are used in industrial and pressure pipeline applications where their special corrosion and /or solvent resistance and higher temperature performance are needed and higher cost justified. In drinking water pipelines, epoxy resins are required.

The primary function of the fabric tube is to carry and support the resin until it is in-place in the existing pipe and cured. This requires that the fabric tube withstand installation stresses with a controlled amount of stretch but with enough flexibility to dimple at side connections and expand to fit against existing pipeline irregularities. The fabric tube material can be woven or non-woven with the most common material being a non-woven, needled felt. Some processes include fabric tube reinforcement of glass fibers or Kevlar®. Impermeable plastic coatings are commonly used on the exterior and/or interior of the fabric tube to protect the resin during installation. The layers of the fabric tube can be seamless, as with some woven materials, or longitudinally joined with stitching or heat bonding.

TABLE 1--Comparison of different trenchless pipeline rehabilitation methods[4]

Method	Diameter Range (in.)	Maximum Installation (ft)	Liner Material	Applications
CIPP:				
Inverted in Place	4-108 (100-2700 mm)	3,000 (900 m)	Thermoset Resin/Fabric Composite	Gravity and Pressure Pipelines
Winched in Place	3-108 (76-2700 mm)	500 (150 m)	Thermoset Resin/Fabric Composite	Gravity and Pressure Pipelines
Sliplining:				
Segmental	4-158 (100-4000 mm)	1,000 (300 m)	PE, PP, PVC, GRP (-EP & -UP)	Gravity and Pressure Pipelines
Continuous	4-63 (100-1600 mm)	1,000 (300 m)	PE, PP, PE/EPDM, PVC	Gravity and Pressure Pipelines
Spiral Wound	4-100 (100-2500 mm)	1,000 (300 m)	PE, PVC, PP, PVDF	Gravity Pipelines Only
In-Line Replacement:				
Pipe Displacement	4-24 (100-600 mm)	750 (230 m)	PE, PP, PVC, GRP	Gravity and Pressure Pipelines
Pipe Removal	up to 36 (900 mm)	300 (100 m)	PE, PVC, PP, GRP	Gravity and Pressure Pipelines
Deformed and Reshaped:				
Fold & Formed	2.5-24 (51-600 mm)	1,500 (450 m)	HDPE, PVC	Gravity and Pressure Pipelines
Drawdown	3-24 (62-600 mm)	1,000 (320 m)	HDPE, MDPE	Gravity and Pressure Pipelines
Rolldown	3-24 (62-600 mm)	1,000 (320 m)	HDPE, MDPE	Gravity and Pressure Pipelines
Point Source Repair:				
Grouting, Link-Seal, Point CIPP	4-24 (100-600 mm)	N/A	Grouting, Special Sleeves, Point CIPP, etc.	Any
Spray-on-Lining	3-180 (76-4500 mm)	500 (150 m)	Epoxy Resins or Cement Mortar	Gravity and Pressure Pipelines
Manhole Rehabilitation	Any	N/A	Spray-On Lining, PVC, CIPP	Sewer Manholes

[4]Definition of Acronyms:
PE - Polyethylene
PP - Polypropylene
EPDM - Ethylene Polypelene Diene Monomer
PVC - Poly-Vinyl-Chloride
PVDF - Poly-Vinylidene Chloride
GRP - Glassfiber Reinforced Polyester
HDPE - High Density Polyethylene
MDPE - Medium Density Polyethylene

CIPP	Sliplining	In-line Replacement	Deformed & Reshaped	Point Source Repair	Manhole Rehabilitation
Inverted in Place - Insituform - Paltem *Winched in Place* - Spiniello KM-Inliner - Inliner U.S.A. - Superliner - Paltem SZ - Eco-Liner - RoboLiner	*Continuous* - Phillips Driscopipe - Spirolite *Segmental* - Hobas - Lamson Vylon Pipe - Duratron *Spiral Wound* - Danby - Rib-Loc - Expanda Pipe	*Pipe Displacement* - PIM - XPANDIT - Grundocrack - Expand-A-Line - Consplit - TRS *Pipe Removal* - Iseki - Soltau - Herrenknecht - Horizontal Holes	*Fold & Formed* - NuPipe - U-Liner - Am-Liner - Sure Line *Drawdown* - Swage-Lining *Rolldown* - Subterra	*Grouting* - KA-TE - Sanipor - Avanti - Cues - Amkrete *Link-Seal* - Weko-Seal - Link-Pipe - Econoliner - Part-Liner *Point CIPP* -Paltem Apollo *Spray-On Lining* - Centriline - Thane Coat - Gunite	*CIPP* - Insituform - Fiberline *Spray-On Lining* - Strong - Fosroc - UltraCure - Raven Chemicals *Cast-in-Place* - Permaform - Monoform - ADS *Profile PVC* - Danby

Figure 1 - Various Trenchless Pipeline Rehabilitation Techniques Groupings

Typical CIPP design properties are shown in Table 2.

Table 2--Typical CIPP Design Properties

PROPERTY	ASTM TEST METHOD	POLYESTER psi (kg/cm^2)	VINYLESTER psi (kg/cm^2)	EPOXY psi (kg/cm^2)
Tensile Strength	D638	3,000 (211)	3,000 (211)	4,000 (281)
Flexural Strength	D790	4,500 (316)	4,500 (316)	5,000 (352)
Flexural Modulus	D790	250,000 (17,581)	375,000 (26,371)	300,000 (21,097)

The primary differences between the various CIPP systems are in the composition of the tube, method of resin impregnation (by hand or by vacuum), installation procedure, and curing process.

Method Description

CIPP involves the installation and curing of a lining material inside the existing pipe. First the pipe to be lined is televised and cleaned. The distance between inspection manholes is measured and all damaged sections and service inlets are carefully noted. The order placed with the CIPP manufacturer is based on this data and the product is manufactured for insertion into this particular pipe.

Roots, grease, sediment and debris are removed from the line. Areas of excessive infiltration are sealed using standard grouting procedures and, if necessary, spot repairs are performed prior to liner placement. In case of inversion method, the inversion head usually overcomes external water pressure and incoming water is pushed ahead of inverting tube so all water is expelled and none is trapped. After the pipeline has been cleaned and televised, a flexible tube which has been manufactured and cut according to specific project conditions and impregnated with a hardening material (resin), is inserted through a manhole or another convenient entry point. It is critical that the fabric tube is totally saturated and air evacuated in order to provide consistent finished physical properties required by design. This flexible tube is installed in the existing pipe by winching mechanically or inverting it under air or water pressure. The flexible tube is then forced by water or air pressure against the wall of the pipe to be rehabilitated and cured by hot water or steam. This process creates dimples at the service connections or laterals. The new pipe is cooled down and drained. The ends of the cured pipe are then trimmed forming a smooth, seamless CIPP. A special cutting device or a robot is used in conjunction with a closed-circuit television (CCTV) camera to re-open service connections, locating the dimples visually with the camera.

After installation, the resin-impregnated fabric cures to form a new pipe of slightly smaller inside diameter, but of the same general shape as the original pipe. As previously stated, there are two primary approaches to installing the flexible tube—inverting in place and winching in-place. Specific variations of installation procedures and materials are employed by different manufacturers.

Advantages

Grouting is not normally required due to the tight fit of the CIPP against the existing pipe. The CIPP has no joints and has a very smooth interior surface which typically improves flow capacity despite the slight decrease in diameter. Non-circular shapes can be accommodated often without a decrease in flow capacity. Lining, if inverted, is capable of accommodating bends and existing pipe deformations. Entry is possible via existing manholes or through minor excavations. Remote control internal lateral connection is possible.

Disadvantages

Trained crew and operators with special equipment are required. The tube or hose must be specially constructed for each project. The existing flow must be by-passed during the installation process. Sealing may be required at liner ends or cuts to prevent inflow or outflow. The curing period could be extensive and may take five hours or more. High set-up costs on small projects may reduce the cost effectiveness of this method.

SLIPLINING

Sliplining is one of the oldest methods of trenchless pipeline rehabilitation with plastic pipe. In this method a new pipe of smaller diameter is inserted into the existing pipe and the annulus between the old pipe and new pipe is usually grouted. This method has the merit of simplicity and is relatively inexpensive.

Method Description

Sliplining or pipe insertion is used to rehabilitate pipelines by sliding a flexible liner pipe of slightly smaller diameter into an existing pipeline. This is done by pulling or pushing new pipe into a deteriorated pipeline. This system is used where there is no obstruction inside the old pipe and its dimensions are in good shape; therefore, it can be used on all types of existing pipes that are not excessively deteriorated. The liner can form a continuous, water-tight pipe within the existing pipe after installation. The annular space between the old and new pipe is usually filled with a grouting material. Sliplining is not considered to be a specialist operation and does not require sophisticated equipment. The main disadvantage is a substantial loss of cross-sectional area. However, a loss of cross-sectional area does not directly relate to loss of hydraulic capacity, especially for larger diameters of more than 30-36 inches (762-914 mm), as there will be some gain from improved hydraulic characteristics. Where the new pipe has to be laid to an even grade, the use of plastic or metal locators/spacers may be necessary. Spacers also maintain the pipe location during annular grouting to ensure a uniform surrounding. The service connections are then reconnected to the new liner by excavation.

Sliplining can be categorized into three types: continuous, segmental and spiral wound. Each method is discussed below.

Continuous

This method can be used both for structural or non-structural purposes. HDPE solid wall pipe is sliplined into an existing pipeline after the joints are butt-fused. This method requires a relatively short trench, depending on depth and diameter of liner, for an insertion pit. There is typically a reduction of the old pipe cross sectional area. It leaves an annular space, which should be grouted, between old and new pipe. Continuous HDPE requires excavation at each lateral connection to reestablish the connection. This method has high mobilization costs and moderate-to-high overall costs.

Advantages--This is a simple method requiring minimal investment in installation equipment and can be carried out by any pipe contractor with relatively little technical skill. The new pipe has very few joints. Lining is capable of accommodating large-radius bends. Lining can be accomplished via either an insertion pit or a manhole (in the case of sewers); the latter requires a pipe with sufficient axial flexibility. In the case of pressure lines, so-called live insertion is possible (flow is continued through the annular space). Reduction in capacity, after the liner is installed, may be compensated by, for example, increasing the pressure.

Disadvantages--Pipes must be manufactured or pre-assembled into a continuous length. Reduction in capacity may be significant, depending on liner pipe diameter and wall thickness. For optimum performance in the case of gravity lines, annular space should be grouted. Only circular cross sections are possible. It is less cost effective where deep external lateral connection is required.

Segmental

This method involves the use of short sections of pipe that incorporate a flush sleeve joint. A number of plastic pipe products, such as PVC and PE, fiberglass reinforced polyester, which include short-length sections with a variety of proprietary joints, have also been specially developed for sliplining sewers.

Advantages--This is a relatively simple method with low investment in installation equipment. An insertion pit may be avoided if pipe sections will fit in a manhole. By applying custom-made shaped pipes (oval, obovate) the pipe capacity reduction can be kept to a minimum. This method requires relatively little technical skill and can be carried out by any pipe contractor or by the municipality itself.

Disadvantages--The annular space between the old and new pipe generally must be grouted. The reduction in pipeline capacity may be significant. There will be many joints in the new pipe. Some pipe materials are easily damaged during the installation. The laterals may need to be connected externally.

Spiral Wound

This technique is based on forming a pipe in-situ by using PVC, PE or PP-ribbed profiles with interlocking edges. The liner is fed through a machine to form a smooth-bore spiral-wound pipe. The winding machine has been modified, enabling it to work from the bottom of a manhole and directly feed a liner into an existing pipe. This method can be used for either structural or non-structural purposes, depending on the type of grouting.

There are three variations of this method: one fabricates a pipe in the manhole by helically winding a continuous PVC fabric; the second, used for larger diameters (over 36 inches or 900 mm), uses preformed panels inserted in-place in the existing pipeline; the third, is used to line oviform man-entry pipes, 36 inches (900 mm) diameter and larger. Excavation is not required for this process. House connections and laterals are reconnected by local excavations, but a remote-controlled cutter is being developed. This method leaves an annular space which must be grouted. It has low mobilization costs and low-to-moderate overall costs.

Advantages--Lining pipe may be formed on-site by spirally winding a strip. Access via manholes is possible. For man-entry pipes, the lining is capable of accommodating large radius bends. No pipe storage on-site is necessary. Within limits, any diameter can be selected within range of the winding machine.

Disadvantages--This method requires trained personnel to operate the winding equipment. Continuous fusion, solvent-welded or mechanical joint is required. Reduction in capacity may be significant. Requires grouting of annulus.

IN-LINE REPLACEMENT

When pipelines are found to be structurally failing and inadequate in capacity, then in-line replacement should be considered. This is typically the most expensive method of trenchless pipeline rehabilitation. However, the money spent is a capital investment since a completely new pipe is installed and often it is very cost effective when compared to open-trench method. There are two categories representing in-line replacement or trenchless method of replacing of an existing pipeline: pipe displacement and pipe removal.

Pipe Displacement

Pipe displacement, also known as pipe bursting, is a technique for breaking out the old pipe by use of radial forces from inside the old pipe. The fragments are forced outward into the soil and a new pipe (usually a HDPE) is pulled into the bore formed by the bursting device.

Method Description--The equipment uses a standard pneumatic mole with a special pointed shield. Twin hydraulically operated breaker arms are attached in the front of the tool. These cracker arms are remotely operated and can exert pressure to overcome difficult joints or surroundings. A standard air compressor is also needed to provide power to the pneumatic mole, and a hydraulic winch is required as well.

Advantages--This method creates no reduction in capacity; in fact, increases in capacity are possible. The new pipe installation can be accomplished by means of an insertion pit (for continuous length) or a manhole (for discrete pipes). If the existing pipe is structurally damaged, the old pipe can be burst or cut (hydraulically or pneumatically) by radial jacking forces and a new pipe pushed or pulled behind.

Disadvantages--Laterals have to be disconnected beforehand and lateral connection carried out by excavation. There is a risk of damaging nearby utilities and building constructions. A crew of trained operators is needed. For replacing metal pipe systems, a sleeve pipe is usually required to protect the inserted pipe lining from damage. Loading can cause changes to the surrounding soil. Existing pipe has to be friable such as clay, cast iron or concrete (not reinforced).

Pipe Removal

The development of microtunneling machines in the late 1980's with the capability of crushing rocks and stones has led to their use in excavating existing pipelines for replacement. The remote-controlled microtunneling equipment is slightly modified for pipe removal systems. The various elements of the system include a modified shield, a jacking unit, a control console and a spoil removal system.

Method Description--A jacking pit of approximately 10 to 20 ft (3 to 6 m) in length (or diameter) and a smaller receiving pit are required. As the microtunnel boring machine is driven forward (excavating the old pipe and surrounding ground), the defective pipe is destroyed and the debris removed while the new pipe sections (usually glassfiber reinforced polyester), with equal or larger outside diameter, is simultaneously pushed up behind the advancing shield.

Advantages--The new pipe can be installed with relatively little surface disruption. The new pipe is capable of accommodating large radius bends. This method can be used to upsize the old pipe. The main advantage of this system is its ability to work in difficult ground conditions, particularly under groundwater. Another advantage is its ability to replace a pipe which is badly out of line and grade, since close control of line and level can be achieved for the new pipe.

Disadvantages--A shaft or insertion pit is required. Obstacles such as manmade structure or roots may cause problems. This method requires a high degree of skill, including a specialized crew with special equipment.

DEFORMED & RESHAPED

Deformed and Reshaped technology originated as the result of a five-year research and development project in France. The first U.S. installation was apparently in Toms River, New Jersey, around September of 1988, using an 8-inch (200-mm) diameter pipe.

Method Description

This type of trenchless pipeline rehabilitation temporarily reduces the cross-sectional area of the new pipe before it is installed, then expands it to its original size and shape after placement to provide a close fit with the existing pipe. Lining pipe first is reduced in size (on-site or in the manufacturing plant), then inserted and finally reverted by heat/pressure or naturally. There are three versions of this approach: Fold and Formed, Drawdown and Rolldown.

Advantages--The reported reduction in pipeline capacity is minimal, if any. Lining can be accomplished via either an insertion pit or a manhole (in the case of sewers), the latter requiring a pipe with sufficient axial flexibility. In this method, no grouting is required. The lining is capable of accommodating large radius bends. The liner can be installed in long lengths, with few or no joints.

Disadvantages--Possible structural damage (collapse/misalignment) to existing piping can cause problems. Lateral connections may be difficult. Different methods vary substantially in required degree of expertise.

Fold and Formed

Method Description--The fold and formed method can be used for both structural and nonstructural purposes. The method uses a jointless extruded PVC or HPDE pipe which is folded or deformed to reduce the cross-sectional area. The folded pipe is mechanically pulled into the existing pipeline, then formed to the shape of the existing pipeline using heat, pressure, and in case of NuPipe and AM-Liner, a mechanized rounding device. Service connections are reinstated internally using a remote-controlled robotics cutter.

In this method it is essential to monitor the existing pipeline by use of a closed circuit television camera (CCTV). In most fold and formed technologies, protruding taps must be reduced to less than 10% of the existing pipe diameter. The taps will normally be reduced by the remote-controlled cutting device. This technique has low mobilization costs and low to moderate overall costs.

Advantages--This system is seamless and jointless, providing excellent flow characteristics and eliminating infiltration at the joints. The installation is rapid. Excavation is not required for this process. Grouting is not required. This method leaves a very small annular space, if any.

Disadvantages--This method has limitations on pipe diameter and length. For example, NuPipe is typically used for rehabilitating lines of 4-inch (100-mm) to 12-inch (300-mm) diameter and up to 600 feet (180 meters) in length and U-Liner is used for 4 to 18-inch (100 to 450 mm) in diameters; however, lengths can vary according to specific job parameters. Lengths are limited as a result of the required pull-in forces and the required heating capacity to thoroughly heat the entire length of the pipe. This system also has limitations on oval or odd shapes of the old pipe, diameter variations, possible joint settlements and the old pipe not being straight.

Drawdown

This method is used to draw the new pipe, after it is heated, through a die to reduce its diameter. Overall diameter reduction ranges from around 20% for a 4-inch (100-mm) diameter pipe, to 7% for a 24-inch (600-mm) pipe.

This method can be used for both structural and non-structural purposes. It inserts an HDPE solid wall pipe into an existing pipeline after joints are butt-fused and the HDPE pipe is swaged down in diameter. Diameter reduction depends on the memory of the polymeric chain structure of medium—and high density polyethylene pipe. Compressing the pipe temporarily crushes the chain structure, allowing the pipe to be reduced in diameter and later reverted to its original size without affecting performance. Pipes of 3 to 24 inches (76 to 610 mm) in diameter can be installed utilizing this method. This system has high mobilization costs, but low-to-moderate overall costs.

Method Description--Long, continuous lengths of the new pipe are pulled into the existing pipe. Then pressure is applied to the inside of the pipe to speed up the reversion process. The new pipe in its reverted form usually fits closely to the old pipe wall, and no annular space remains.

Advantages--An excavation is required, but it will be shorter than required for regular sliplining. House connections are reconnected by a remote-controlled cutter without excavation. Since the liner forms tightly against the wall of the original pipe, it leaves no annular space, eliminating the need for annulus grouting.

Disadvantages--Temporary excavation is necessary to install the liner, since existing manholes do not provide enough space. The liner may not invert if there are diameter variations in the existing pipe. There is a possibility of stress cracking due to elongation of the liner during the insertion. The liner relies on the existing pipe for installation support.

Rolldown

The Rolldown system is similar to drawdown except that the new pipe diameter is reduced for insertion by running the new pipe through a cold rolling machine. This rearranges the long chain structure of the plastic pipe to produce a smaller diameter pipe with thicker walls and minimal elongation.

Method Description--In this system, sections of standard medium density or high density polyethylene (MDPE or HDPE) are butt-fused into appropriate lengths. Then the pipe is cold-rolled on site to reduce its diameter to allow insertion. After the liner pipe has been inserted into the old pipe, the liner is pressurized to restore the pipe to its original size, resulting in the tight fit inside the pipe.

Advantages--Rolldown has an advantage over drawdown in that the treated pipe will remain at the reduced diameter for a considerably long period of time. The rolldown process also does not rely on mechanical means to prevent the new pipe from reverting to its original size prior to insertion.

Disadvantages--Temporary excavation is necessary to install the new pipe, since existing manholes do not provide enough space. The liner may not invert if there are diameter variations in the existing pipe. The liner relies on the existing pipe for installation support.

POINT SOURCE REPAIR

When local defects are found in a structurally sound pipeline, due to cracking or joint failure, point source repairs are considered. There are two variations of this method: grouting-link seal and spray-on-lining.

Grouting-Link Seal

Systems are available for remote-controlled resin injection to seal localized defects in the range of 4-24 inches (100 to 600 mm) in diameter. The new point source repair devices (such as grouting sleeves) are also used to address basically four problems. The first objective is to keep the loose and separated pieces of un-reinforced old pipe aligned to maintain a load bearing equivalent to a masonry arch. The second purpose is to provide added structural capacity or support to assist the damaged pipes to sustain structural loads. The third goal is to provides a seal against infiltration and exfiltration. The forth objective is to replace missing pipe sections.

Advantages--Different geometric configuration can be accommodated. Oval pipes and even box sections can be repaired.

Disadvantages--This method may result in reduction in flow capacity of the pipe.

Spray-On-Lining

The spraying of a thin mortar lining or a resin coating onto pipes is a well established technique. Such methods can provide improved hydraulic characteristics and corrosion protection. However, they may not enhance the structural integrity of the line and have little value in sealing joints or leaks. For man-entry pipes, structural reinforced sprayed mortars (shotcrete or gunite) are effective and widely used for rehabilitating pressure pipes and gravity sewers.

For non-man entry pipes, the lining is sprayed directly onto pipe walls using a remote-controlled travelling sprayer. The lining materials include concrete sealers, epoxy, polyester, silicone, vinylester, and polyurethane.

Advantages-- Because liner material is sprayed on as a viscous liquid, it cannot block the lateral openings.

Disadvantages--The system requires full flow diversion and cleaning of the pipe before application.

SEWER MANHOLE REHABILITATION

Sewer manholes require rehabilitation to prevent surface water inflow and groundwater infiltration, to repair structural damage and to protect surfaces from damage of corrosive substances. When rehabilitation methods will not solve the problems cost-effectively, manhole replacement should be considered. Selection of a particular rehabilitation method should consider the type of problems, physical characteristics of the structure, location, condition, age and type of original construction. Extent of successful manhole rehabilitation experiences and cost should also be considered.

The sewer manhole rehabilitation can be divided into the following methods:

- Spray-on lining
- Cast-in-place
- Cured-in-place
- Profile PVC

For more information on manhole rehabilitation techniques and for a list of advantages and disadvantages of each technique, the reader is referred to U.S. Environmental Protection Agency's Handbook of Sewer System [4].

CONCLUSIONS

This report described some of the features of relevant rehabilitation methods. New and potentially viable methods were also mentioned. The design engineer and utility owner should be aware of these methods and others as their technologies develop and advantages and disadvantages of each. The choice of methods depends on the physical condition of the pipeline system components, such as, pipeline segments, manholes, and service connections; and the nature of the problem or problems to be solved.

ACKNOWLEDGMENT

As stated previously, this study is part of a state-of-the-art review which is being prepared by the Trenchless Technology Center (TTC) for the U.S. Army Corps of Engineers' Construction Productivity Advancement Research (CPAR) program. Permission was granted by the Chief of Engineers to publish this information. Under this CPAR program, methods and materials in trenchless technology will be studied and several field and laboratory tests will be conducted. Readers are encouraged to send their comments or information on new trenchless rehabilitation products and methods to authors at the Trenchless Technology Center, Louisiana Tech University, P.O. Box 10348, Ruston, LA 71272, Phone (318)257-3204, Fax (318)257-2562.

REFERENCES

[1] Schrock, B.J., "Pipeline Rehabilitation: A Guide to a Growing Market," The National Utility Contractor, NUCA, Arlington, Virginia, 1992, 16(8), 14-21.

[2] Erdos, L.I., "Rehabilitation of Urban Pipelines," Proceedings of the 1991 National Conference and Symposium on Water Resources Planning and Management and Water Resources, New Orleans, Louisiana, 1991.

[3] Southwest Contractor, February 1989.

[4] "Handbook of Sewer System Infrastructure Analysis and Rehabilitation," U.S. Environmental Protection Agency (EPA), Office of Research and Development, Cincinnati, Ohio, 1991.

Mohammad Najafi[1] and David T. Iseley[2]

EVALUATION OF PVC PIPE FOR MICROTUNNELING

REFERENCE: Najafi, M. and Iseley, D.T., "Evaluation of PVC Pipe for Microtunneling," Buried Plastic Pipe Technology: 2nd Volume, ASTM STP 1222, Dave Eckstein, Ed., American Society for Testing and Materials, Philadelphia, 1994.

ABSTRACT: Polyvinyl chloride (PVC) pipe is a suitable product for sewer system construction. However, in the past, due to axial thrust load limitations and lack of a suitable joint it was not possible to use PVC pipe in microtunneling. In June 1992, a full-scale field test was conducted at Louisiana Tech University and a sewer PVC profile pipe was successfully installed with a new microtunneling propulsion system. The 24-inch plastic pipe with smooth joint used for this evaluation program was inspected after installation and was monitored for one year for effects of construction loads and long-term vertical soil loads. The details and results of this field testing will be discussed in this paper.

KEY WORDS: polyvinyl chloride (PVC) pipe, microtunneling, trenchless technology, pipe design, underground infrastructure, sewer pipe installation

---

INTRODUCTION

The demand for installing new underground utility systems in congested areas with existing utility lines has increased the necessity for innovative systems to go underneath in-place facilities. Environmental concerns, social (indirect) costs, new safety regulations, difficult underground conditions (existence of natural or artificial obstructions, high water table, etc.) and new developments in equipment have increased the demand for trenchless technology. Trenchless technology methods include all methods of installing or rehabilitating underground utility systems with minimum disruption of the surface.

Microtunneling is one of the many methods and the fastest growing method among a family of trenchless techniques. Microtunneling methods are usually for utility lines that are less than or equal to 900 millimeters (36 inches) in diameter and, therefore, require non-man

---

[1]Assistant Professor, Department of Engineering Technology, Missouri Western State College, Saint Joseph, MO 64507.

[2]Associate Professor of Civil Engineering and Director of the Trenchless Technology Center (TTC), Louisiana Tech University, Ruston, LA 71272.

entry tunnel boring machines. The same method, however, can be used for diameters larger than 900 millimeters (36 inches). Microtunneling boring machines (MTBM's) are laser-guided, remotely-controlled, and have the capability to install pipelines on precise line and grade. This method can be used to install pipes in virtually any type of ground conditions, up to 45 meters (150 feet) below the ground surface, and up to 225 meters (750 feet) in length from the drive to the reception shaft [1].

During recent years, there has been remarkable progress in development of new microtunneling equipment. These developments have produced new improvements in jacking force capacity, length of drive, steering capabilities and other advancements. However, the types of usable pipes for microtunneling have been limited. This is because in the current microtunneling methods, the jacking force is directly exerted on the pipe, making it necessary for the pipes to have a sufficient strength to resist that force. This negatively impacts microtunneling methods when compared with open-trench construction, where a wide variety of pipe materials with excellent corrosion and hydraulic characteristics but low end-bearing strength, such as polyvinyl chloride (PVC), can be used.

In order to enable microtunneling systems to use a wide variety of less expensive, thin-wall pipes, an innovative microtunneling system, called LLB (Laying pipes of Low Bearing force) was developed by Kidoh Construction Company of Osaka, Japan. Since a major portion of costs for microtunneling operations is the cost of thick-wall pipe to resist axial jacking loads, microtunneling projects would become more cost competitive with open-trench construction if less expensive pipes can be used.

PIPE JACKING

Requirements of Jacking Pipe

In general, pipe used for microtunneling must be round, have a smooth, uniform outer surface, and with watertight joints that also allow for easy connections between pipes. Pipe lengths must be within specified tolerances and pipe ends must be square and smooth so that jacking loads are evenly distributed around the entire pipe joint and such that point loads will not occur when the pipe is jacked in a reasonably straight alignment. Pipe used for pipe jacking is capable of withstanding all forces that will be imposed by the process of installation, as well as the final in-place loading conditions. The driving ends of the pipe and intermediate joints are protected against damage as specified by the manufacturer. The detailed method proposed to cushion and distribute the jacking forces is specified for each particular pipe material.

Any pipe showing signs of failure may be required to be jacked through to the reception shaft and removed. The pipe manufacturer's design jacking loads should not be exceeded during the installation process. As a factor of safety, the ultimate axial compressive strength of the pipe must be a minimum of 2.5 times the maximum design jacking load of the pipe. At present time the following pipe materials specially manufactured for microtunneling operations are available: 1) glassfiber reinforced polyester (GRP) pipe, 2) reinforced concrete pipe, 3) vitrified clay pipe, 4) steel pipe, 5) resin concrete pipe, and 6) ductile iron pipe. This project is the first time in the United States that a PVC sewer pipe is installed with microtunneling.

Advantages of PVC Pipe

Polyvinyl chloride (PVC) pipe was first produced and installed on a very limited basis in Germany in mid 1930's. The use of PVC pipe as a sewer pipe in the United States started in 1960's. Today PVC pipe commands a large share of the world market, including the United States. In 1991, PVC was the largest volume of plastic used in the construction industry (45 percent of all plastic usage), mainly for pipe and tubing [2]. About 90 percent (based on the weight) of all plastic pressure-water pipe is PVC and almost 100 percent of plastic sewer pipe is PVC [3].

Some of the advantages of PVC pipes for sewer system construction include the following: 1) lightweight and easy to handle, 2) good impact resistance and toughness, 3) resistance to a wide range of corrosive environments found in sewage and soil, 4) good hydraulic flow characteristics, 5) easy to work with, 6) economical, 7) durable, 8) availability of different joint systems which are reliable against leakage, 9) excellent abrasive resistance, 10) excellent dimensional control, and 11) not biologically degradable. The use of PVC sewer pipe has decreased infiltration and exfiltration and accompanying tree-root problems. The surfaces of the pipe are very smooth and resist buildup of deposited materials and other solids.

Compared to other types of pipe materials available for microtunneling such as reinforced concrete, steel and glassfiber reinforced plastic mortar pipe (GRP), PVC sewer pipe has less unit cost. However, in the past, the following factors were the main obstacles to the utilization of PVC pipe in microtunneling:

1. axial thrust load limitations,
2. higher cost of thick-wall PVC pipe to resist the thrust load, and
3. lack of a suitable joint compatible for microtunneling.

Forces Acting on the Pipe

In microtunneling methods the jacking force required to push a pipe theoretically consists of the penetration resistance of the boring and steering head and the frictional resistance of the product pipe. In accordance with Figure 1, the jacking force can be expressed in the following equation [4]:

$$FM = F0 + \sum FP \quad (1)$$

where:

FM = Total jacking (thrusting) force, (ton),
F0 = Resistance of the MTBM (penetration resistance), (ton), and
FP = Frictional resistance (loads acting in the direction of pipe axis) (ton).

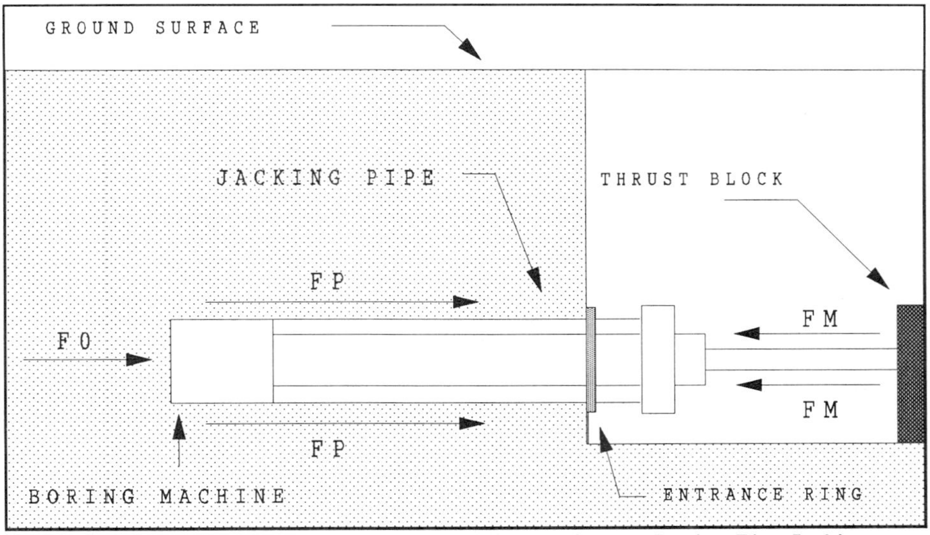

Figure 1 - Frictional and Face (Penetration) Resistance During Pipe Jacking

The jacking force in conventional microtunneling systems must be axially applied to the pipe from the main jacking station. Moreover, it must be ensured that the resulting stresses stay below the allowable value in all jacking pipe sections. Proceeding from the main jacking station, the jacking force decreases, pipe by pipe, by the amount of the given frictional resistance. Therefore, the highest jacking force in conventional microtunneling methods theoretically occurs in the pipe closest to the main jacking station just before the far end of the pipe arrives at the receiving pit. However in the LLB system, the maximum jacking force on the pipe will occur at the pipe sections which hold the gripper system.

The amount of jacking force required is governed by the soil type and its characteristics such as soil density and water content (location of water table); project characteristics such as height of cover, size of overcut, lubrication, overburden loads, time and jacking distance; and pipe characteristics such as pipe size, dimensional consistency, weight, resiliency, absorbency and smoothness of its outer surface. On the other hand, the amount of jacking force which can be applied in any microtunneling project is limited to a great extent to the strength of the pipe material and cross-sectional area of the jacking pipe at the minimum cross-sectional area, extent of eccentricity of the resultant jacking force, capacity of jacking equipment, and load-bearing capacity of the thrust block. The design parameters—such as the jacking rate and the interaction of the soil and/or water pressure at the face—are also important in determining the design jacking force. The oversize cut or use of lubricants may decrease the jacking force up to 30% or more. On the other hand, occurrence of any unexpected obstructions, such as existence of boulders or restraint of pipes due to steering errors, can bring about a sudden increase in the jacking force. Consequently, it is desirable to install pressure relief valves at the drive pit and provide indicators on the control panel to ensure that the allowable jacking force is not exceeded.

THE LLB SYSTEM

The Description of the LLB System

In the LLB system, the requirement for the pipe to have a high axial load capacity has been removed with a method of transmission of the thrust force by the addition of a liner casing and gripper system. This system transfers the face resistance of the machine to the liner casing temporary inserted in the product pipe. Also, the circumferential frictional resistance of the product pipe is transferre to the gripper systems which are installed at certain intervals along the length of the pipe. These gripper systems expand with air pressure and connect the liner casing with the inner surface of the pipes to transmit the thrust force of the liner casing to the pipes. Therefore, the maximum thrust force exerted to the product pipes is equal to the circumferential frictional resistance of the portion of product pipes between the gripper locations. Consequently, with utilization of the LLB system, pipes with relatively low compressive strength can be installed with microtunneling methods.

In addition to reducing the thrust force on the plastic pipe, the gripper system also has the following characteristics:

1. There is minimal possibility of damaging pipes because the contact is made by pneumatically inflated rubber tubes.
2. Conventional microtunneling methods require product pipes of special wall thicknesses. The LLB system does not have these requirements.
3. The grippers can be installed at desired locations depending on the level of circumferential resistance of the pipes.
4. Long-distance thrusting with minimal restriction to pipe compressive strength is possible with the LLB system.

Another advantage of the LLB system is that it can be used with both auger and slurry shield microtunneling equipment. For the auger type, its screw conveyer's casing can also be used as a propulsion shaft for transmitting the thrusting force. For the slurry type, as stated previously, a steel pipe is utilized as the propulsion shaft, and is inserted through the product pipe before jacking.

Location of Grippers

The following equations provided the number of grippers required for this testing program. The number of required grippers or number of pipe sections which can be jacked directly, without grippers, is related to the load bearing capacity of gripper section or pipe section and frictional resistance of the pipe. Figure 2 illustrates the total thrusting distance (L) to be divided into pipe sections jacked by the grippers (G1, G2, to G_n) and pipe sections jacked directly (DI). The total number of grippers required for each soil section equals to the allowable jacking capacity of the pipe or the gripper system divided by the frictional resistance of the pipe, or FP, as stated in equation 1. Therefore:

$$NG = \frac{GC}{R \times S \times L} = \frac{GC}{FP} \quad (2)$$

$$NP = \frac{PA}{R \times S \times L} = \frac{PA}{FP} \quad (3)$$

where:

NG = Number of PVC pipes to be jacked by one gripper,
NP = Number of PVC pipes to be jacked directly,
PA = Allowable jacking force capacity of PVC pipe, (ton), and
GC = Allowable jacking force capacity of one gripper, (ton).

Note: In order not to exceed the bearing capacity of pipe section or gripper system, in the above equations, the one which has the lower value between GC or PA should be used. In this evaluation program, the value of PA (9.9 tons) is lower than that of GC (40 tons). Therefore, the value of PA was used.

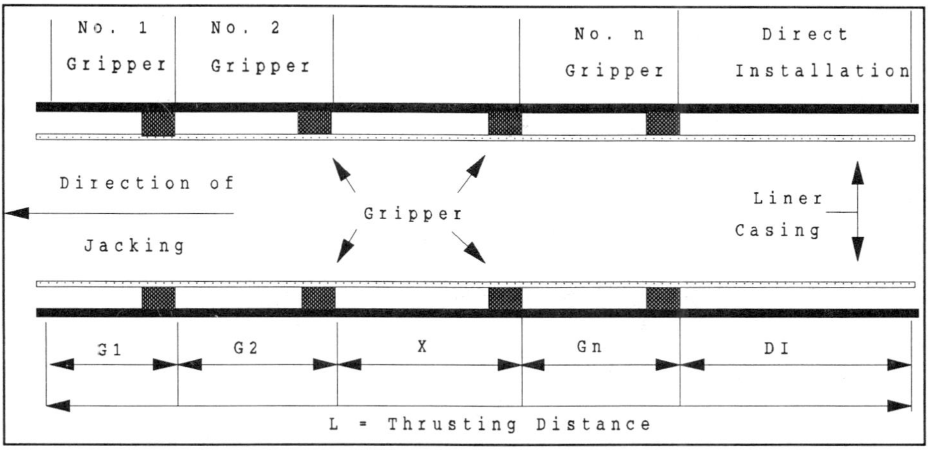

Figure 2 - Estimating Location of Grippers

PVC Pipe Used for the Project

After reviewing the available options of different PVC products for sewer construction, Vylon PVC Sewer Pipe manufactured by Lamson Vylon Pipe, Cleveland, Ohio, was selected for this evaluation program. The Vylon pipe provided suitable characteristics and eliminated both of the major obstacles other PVC pipes experience when used for microtunneling, that is, cost and suitable joint.

The Vylon pipe utilizes a new joint system developed by Lamson Vylon Pipe for trenchless technology. The joint provides a smooth outside and inside transition from one pipe section to another, making the pipe suitable for both sliplining and microtunneling applications. This connection permits the pipe and joint system to mate up with the MTBM. Additionally, an air-tight seal is formed at the joint with a multi-fin gasket wrapped around a fiberglass insert ring. Compared to PVC solid wall pipe, this pipe is more economical due to utilizing a profile wall. Figure 3 illustrates the dimensions and details of the PVC pipe and jointing system.

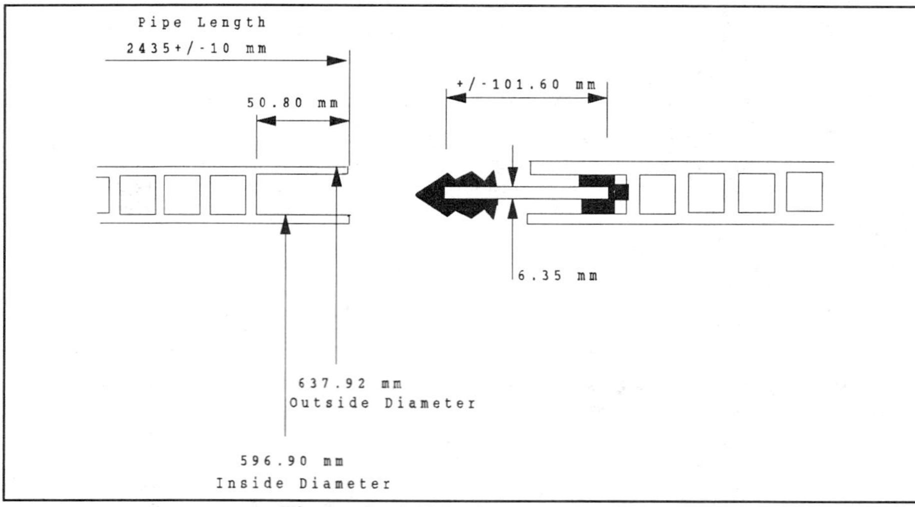

Figure 3 - PVC Pipe Wall Section

The pipe sections selected for use in the project were each 2.438 meters (8 feet) in length and weighed approximately 29.76 kg/m (20 pounds per foot). It was decided early in the planning stage that a tolerance of 150 millimeters (5.9 inches) would be acceptable for the pipe length in order to correspond to the length of inner steel liner casing. Should the total tolerance have exceeded 150 millimeters (5.9 inches), then special spacers would be required at the pipe thrust plate on the thrust shaft side. Kidoh engineers indicated that the desired pipe outside diameter should be 645 mm ± 10-20 mm (25.4 inches ± 0.4-0.8 inches) with the preferred range being 635-655 millimeters (25-25.8 inches) to be consistent with the MTBM.

The Vylon PVC sewer pipe is manufactured according to ASTM F-794 for pipe requirements and ASTM D-3212 for joint requirements. The modulus of elasticity of the pipe is 36,556 kgf/cm^2 (520,000 psi) which is important in order to determine the amount of pipe diameter expansion due to the internal pressure of the gripper. The pipe has a stiffness of 3.2 kgf/cm^2 (46 psi) according to ASTM D-2412. The pipe at the time of testing was under development and had an axial compressive load capacity of 9.9 tons (22,000 lb). The outside diameter of the pipe is 638 millimeters (25.1 inches), the inside diameter 597 millimeters (23.50 inches), and the wall thickness is 20 millimeters (0.80 inches).

TESTING OPERATION

During the month of May and the first half of June of 1992 a test bed was constructed on the Louisiana Tech University campus. The test bed constructed was 58.52 meters (192 feet) in length. A 43.89-meter (144-foot) section of the test bed was excavated, 2.13 meters (7 feet) wide and 3.05 meters (10 feet) deep, and backfilled with equal sections of sand, silt and clayey gravel. An additional 14.6-meter (48-foot) section of the test bed was undisturbed stiff clay. The length of each section of the test bed was chosen to be the equivalent of six 2.438-meter (8-foot) pipe sections. For slurry microtunneling methods a minimum of 2 meters (6.6 feet) overburden on top of the pipe is required. This requirement is to provide enough pressure to prevent slurry loss. In order to prevent side effects, it was also necessary to have at least 0.34 meter (1.2 feet) select backfill below and 0.5 meters (1.6 feet) on each side of the proposed bore. Figures 4 and 5 present the plan and cross section of the test bed. Table 1 presents a summary of project conditions.

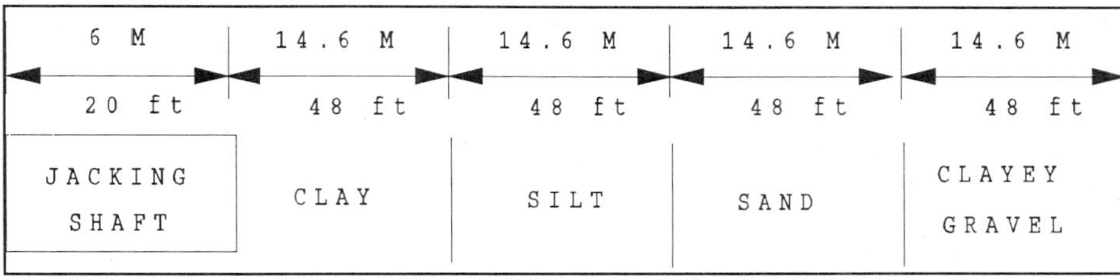

Figure 4 - Test Bed Plan

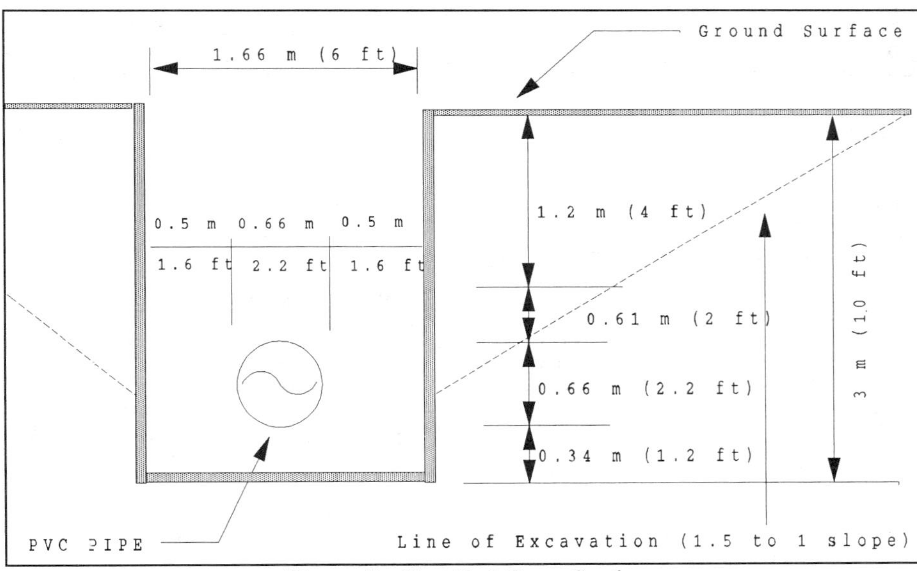

Figure 5 - Test Bed Cross Section

TABLE 1--Summary of project conditions

Description	Quantity
Minimum Depth (from top of the pipe)	1.83 meters (6 feet)
Maximum Depth (from top of the pipe)	2.74 meters (9 feet)
Total Length of Drive	58.52 meters (192 feet)
Diameter of Bore	660 mm (25.98 in.)
Outside Diameter of Pipe	640 mm (25.20 in.)
Size of Overcut (on Radius)	10 mm (0.39 in.)
Inside Diameter of Pipe	599 mm (23.58 in.)
Pipe Thickness	20 mm (0.79 in.)
Soil Dry Unit Weight (Clay)	1,448.21 kg/m^3 (90.4 pcf)
Soil Dry Unit Weight (Silt)	1,746.18 kg/m^3 (109.0 pcf)
Soil Dry Unit Weight (Sand)	1,694.92 kg/m^3 (105.8 pcf)
Soil Dry Unit Weight (Clayey Gravel)	1,890.36 kg/m^3 (118.0 pcf)

TEST RESULTS

The testing operation was successfully completed ahead of schedule on June 22. During this testing operation, for the first time in the United States, approximately 60 meters (200 feet) of 610-mm I.D. (24-inch) PVC sewer pipe was installed with a MTBM. The pipe jacking operation was performed in 8 working days, with an average production rate of 7.62 meters (25 feet) a day. This production rate was excellent considering difficult ground conditions, extra time required for data acquisition which is not normally a part of a typical microtunneling project, and further considering the fact that the operation was experimental and subject to effects of the learning curve.

Significant aspects of this unique testing program were the evaluation of the pipe joint and effects of loading on the plastic pipe. An oil-pressure load meter was located on the pusher ring of the jacking unit. This load meter was used to measure the axial compressive load both at the beginning and end of the pipe section. The data obtained from this load meter was transferred to the control panel to inform the microtunneling operator how well the grippers were working. These data were necessary to keep the maximum jacking force on the PVC pipe below the allowable limits.

The maximum compressive load on the PVC pipe occurred at pipe number 12 (29 meters or 96 feet from the drive pit), when the MTBM was entering the sand zone, and was found to be 3.5 tons (7,700 lb). The maximum jacking force on the thrust block, however, occurred while jacking pipe number 24 and was found to be 41 tons (90,405 lb). Figure 6 compares the actual loads on the PVC pipe and on the thrust block.

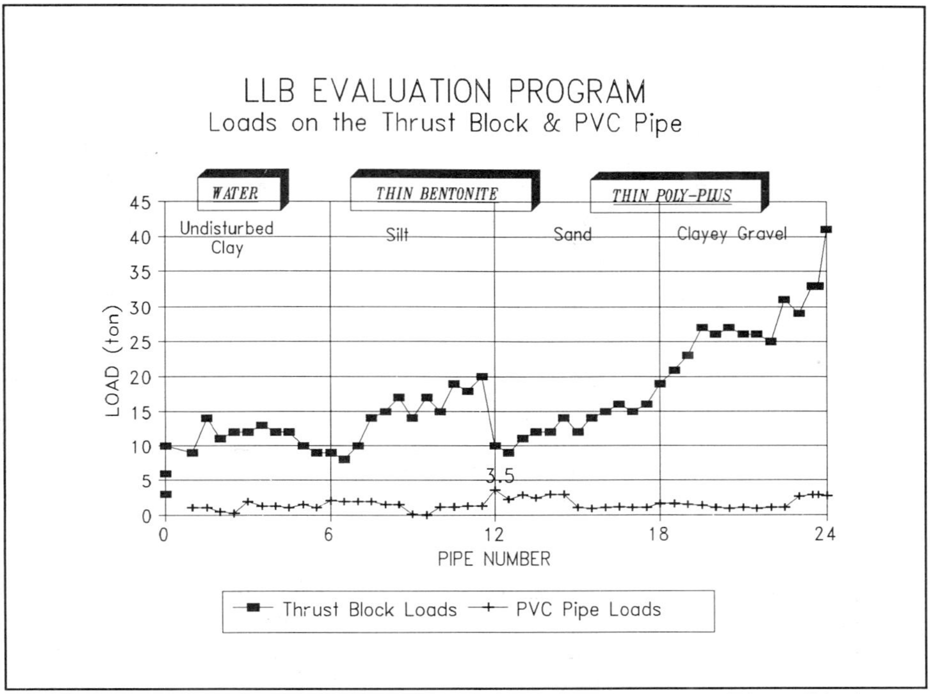

Figure 6 - Comparison of Loads on the PVC pipe and on the Thrust Block

Evaluation of PVC Pipe

After the installation of the pipe was completed, the pipe was air tested up to 0.25 kg/cm^2 (3.5 psi) for a duration of 3 hours. In addition, the horizontal and vertical diameters and lengths of installed PVC pipe were measured according to Figure 7. Table 2 presents a comparison of pipe measurements just before installation and after installation by the LLB system.

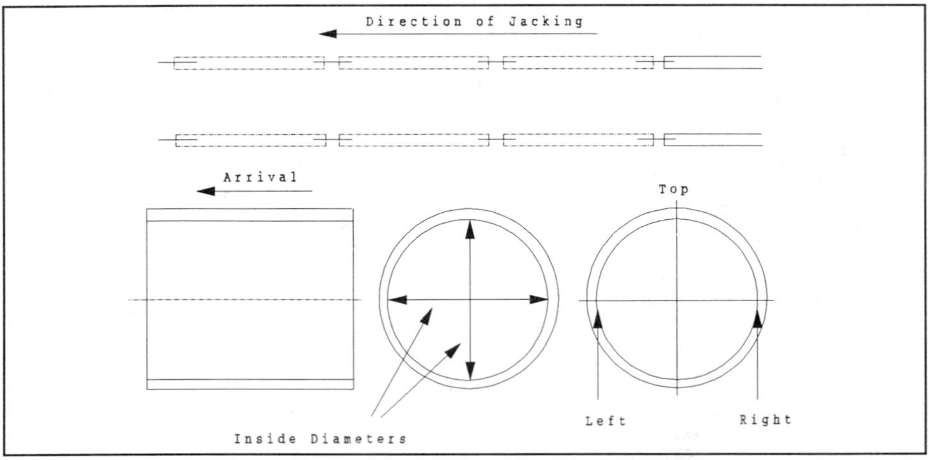

Figure 7 - Measurements of Installed PVC Pipe

TABLE 2--Comparison of Vylon PVC pipe measurements before and after installation

No.	On the Ground (mm)		Location	Pipe Length (mm)			Joint Gap (mm)			I. Dia. (mm)	
	Av. I. Dia	Av. Length		Before Inst.	After Inst.	Difference	Before Inst.	After Inst.	Difference	H. Dia.	V. dia
1			Left Top Right	? ? ?	2435 2432 2433					596	600
2			Left Top Right	2435 2433 2432	2433 2432 2433	-2 -1 1	42 42 44	23 24 19	-19 -18 -25	596	602
3			Left Top Right	2433 2432 2431	2432 2432 2433	-1 0 2	46 45 49	18 21 22	-28 -24 -27	594	602
4			Left Top Right	2430 2431 2431	2432 2428 2433	2 -3 2	37 35 35	13 19 16	-24 -16 -19	600	600
5			Left Top Right	2431 2431 2431	2431 2432 2432	0 1 1	33 31 31	34 36 35	1 5 4	588	598
6			Left Top Right	2431 2431 2431	2433 2432 2432	2 1 1	31 30 31	29 29 31	-2 -1 0	604	600
7			Left Top Right	2430 2432 2429	2431 2430 2430	1 -2 1	32 30 32	25 23 22	-7 -7 -10	598	598
8	598	2417	Left Top Right	2415 2414 2417	2416 2413 2415	1 -1 -2	25 21 21	17 16 9	-8 -5 -12	600	598

TABLE--2 (Continued)

No.	On the Ground (mm)		Location	Pipe Length (mm)			Joint Gap (mm)			I. Dia. (mm)	
	Av. I. Dia	Av. Length		Before Inst.	After Inst.	Difference	Before Inst.	After Inst.	Difference	H. Dia.	V. dia
9	598	2417	Left	2416	2416	0	18	6	-10	596	598
			Top	2416	2412	-4	18	6	-10		
			Right	2417	2417	0	13	4	-9		
10	597	2414	Left	2415	2414	-1	9	-1	-10	598	598
			Top	2413	2415	2	19	11	-8		
			Right	2412	2416	4	22	13	-9		
11	597	2416	Left	2416	2416	0	26	13	-13	596	598
			Top	2413	2414	1	16	3	-13		
			Right	2417	2418	1	20	5	-15		
12	597	2419	Left	2415	2417	2	7	0	-7	598	598
			Top	2419	2417	-2	12	13	1		
			Right	2420	2423	3	19	15	-4		
13	598	2417	Left	2418	2414	-2	20	26	5	598	598
			Top	2413	2415	2	17	19	2		
			Right	?	2424			13			
14	597	2420	Left	2421	2422	1	11	3	-8	602	598
			Top	2420	2420	0	12	10	-2		
			Right	2420	2424	4	25	9	-16		
15	598	2417	Left	2413	2417	4	38	56	18	598	598
			Top	2416	2413	-3	35	57	22		
			Right	2419	2420	1	39	56	17		
16	598	2413	Left	2410	2417	7	25	25	0	600	598
			Top	2412	2410	-2	20	24	4		
			Right	2414	2414	0	13	12	-1		
17	598	2413	Left	2409	2413	4	37	29	-8	598	598
			Top	2410	2413	3	33	28	-7		
			Right	2411	2413	2	28	29	1		
18	599	2418	Left	2418	2420	2	16	3	-13	602	598
			Top	2417	2418	1	21	12	-9		
			Right	2415	2419	4	17	10	-7		
19	598	2414	Right	?	2418			15		598	598
			Top	?	2412			12			
			Left	?	2416			5			
20	597	2413	Right	2410	2414	4	16	10	-6	600	600
			Top	2410	2413	3	18	11	-7		
			Left	2414	2411	-3	25	15	-10		
21	597	2413	Right	2414	2407	-7	29	17	-12	598	596
			Top	2417	2415	-2	21	13	-8		
			Left	2411	2414	3	19	12	-7		
22	597	2417	Right	2417	2415	-2	8	5	-3	600	598
			Top	2417	2418	1	13	8	-5		
			Left	2412	2419	7	28	8	-20		

TABLE--2 (Continued)

No.	On the Ground (mm)		Location	Pipe Length (mm)			Joint Gap (mm)			I. Dia. (mm)	
	Av. I. Dia	Av. Length		Before Inst.	After Inst.	Differ ence	Before Inst.	After Inst.	Differ ence	H. Dia.	V. dia
23	598	2414	Right	2414	2414	0	9	10	1	596	600
			Top	2415	2416	1	11	16	5		
			Left	2414	2415	1	11	20	9		
24	598	2417	Right	2413	2416	3	14	11	-3	598	598
			Top	2416	2417	1	10	13	3		
			Left	2416	2415	-1	10	15	5		

NOTE: The question marks (?) stand for missing data.

The above measurements did not show any significant changes over the original pipe measurements. The monitoring of the pipe for possible deformations in result of long-term soil loading was continued for one year but no pipe deformations were found. These measurements together with theoretical analyses of microtunneling sewer PVC pipe will make a great contribution to the state-of-the-art of buried flexible thermoplastic piping—not just PVC sewer piping. This is the first time that, based on data obtained on a pipe installed with microtunneling these analyses have been conducted.

CONCLUSIONS

The Vylon PVC sewer pipe, in spite of its low bearing strength, was successfully installed by microtunneling with the LLB system. The location of the head of the pipe was measured by surface surveying instruments and found to be within a few millimeters of the target. The 58.5-meter (192-foot) installation was completed in 8 days or an average of 7.6 meters (25 feet) a day. Subsequent evaluation of the PVC pipe showed that the pipe installation was completed without any pipe failure. After completion of installation, the PVC pipe passed the air test. This test shows that the joints were watertight and, had this pipe been installed for a real sewer construction, it would be acceptable for use. Installation of PVC pipe with microtunneling provided an excellent opportunity to monitor the pipe for possible deflections under construction and long-term loadings and known field conditions. At the time of this writing, insignificant deflections have been found in the PVC pipe as a result of long-term loadings. Plans are under progress to investigate effects of surcharge loading on the pipe.

ACKNOWLEDGEMENT

This research project could not have been possible without the financial support of Kidoh Construction Company of Osaka, Japan; Iseki Poly-Tech of Tokyo, Japan; and Lamson Vylon Pipe of Cleveland, Ohio. The authors wish to express their appreciation to these companies and their principles for their participation in this evaluation program.

REFERENCES

[1] Iseley, D.T. and Tanwani, R., Editors, "Trenchless Excavation Construction Equipment and Methods," National Utility Contractors Association (NUCA), Arlington, 1992.

[2] Lawrence, J.R., "Plastics: New Materials of the Century," Engineering News Record, Vol. 229, No. 21, November 23, 1992, pp 3-18.

[3] Moser, A.P. "Buried Pipe Design," McGraw-Hill, Inc., New York, 1990.

[4] Stein, D., Mollers, K. and Bielecki, R. "Microtunneling," Ernst & Sohn, Berlin, Germany, 1989.

Norman Brown [1] and Xici Lu [2]

THE EFFECT OF LOADING RATE ON RAPID CRACK PROPAGATION IN POLYETHYLENE PIPES

REFERENCE: Brown, N. and Lu, X., "The Effect of Loading Rate on Rapid Crack Propagation in Polyethylene Pipes", Buried Plastic Pipe Technology: 2nd Volume, ASTM STP 1222, Dave Eckstein, Ed., American Society for Testing and Materials, Philadelphia, 1994.

ABSTRACT: The effect of loading rate on the production of rapid crack propagation (RCP) was investigated in tensile loaded notched specimens 127 mm long taken from a gas pipe. It was found that the stress intensity to initiate fracture was independent of loading rate, but was slightly greater at 10°C than at 28°C. The initiation of rapid fracture occurred at a critical loading rate and rapid long range fracture (RCP) occurred at a 50% higher loading rate. It was not possible to produce RCP at 28°C with the available equipment. The input energy to initiate rapid fracture was about 10 times the input energy to produce RCP, long range rapid fracture.

KEY WORDS: Rapid fracture, gas pipes, loading rate, K_{IC}, notched specimen, crack propagation.

Introduction

When a pressurized pipe is fractured by a sharp impact, it is possible for the crack to propagate a long distance along the pipe. This can be a catastrophic event and is known as rapid crack propagation (RCP). Very few RCP failures have been observed in polyethylene gas pipes, but as the diameter of the pipe and the pressure are increased the probability of RCP increases. Because the consequences of an RCP failure are very severe, research was initiated by British Gas about 17 years ago. Up to the present time about 10^7 dollars have been spent on RCP testing and research.

The current test methods for observing RCP is to measure the critical pressure above which RCP occurs. Greig [1] at British Gas developed the full scale test where the pipe is pressurized at 0°C and then impacted at one end with a sharp striker. The end of the pipe where fracture is initiated is cooled to a very low temperature in order to introduce a very sharp crack. The full scale test is expensive costing about $20,000/test and has the added feature that the pipe is depressurized as the crack propagates. To meet these objections a smaller scale pipe test was developed by Yayla and Leevers et. al.[2] at

---

[1] Professor and [2] Senior Scientist in Department of Materials Science and Engineering, University of Pennsylvania, Philadelphia, PA 19104.

Imperial College and which costs several thousand dollars for a given pipe material. This test has been adapted by the International Standards Organization (ISO) and is known as the S4, Small Scale Steady State, test. In the S4 test the rate of decompression is reduced by placing a series of baffles inside the pipe. Also the effect of the support of the earth in the full scale test is somewhat duplicated by placing a cage around the outside wall of the pipe. Typical test results from the S4 test are shown in Fig. 1. Note that the transition from no propagation to RCP occurs over a very narrow range of pressure. The critical pressure, P_c, from the British Gas full scale test is about $4 \times P_c$ from the S4 test as a result of the decompression. Kanninen et. al. [3] have analyzed the effects of decompression on P_c.

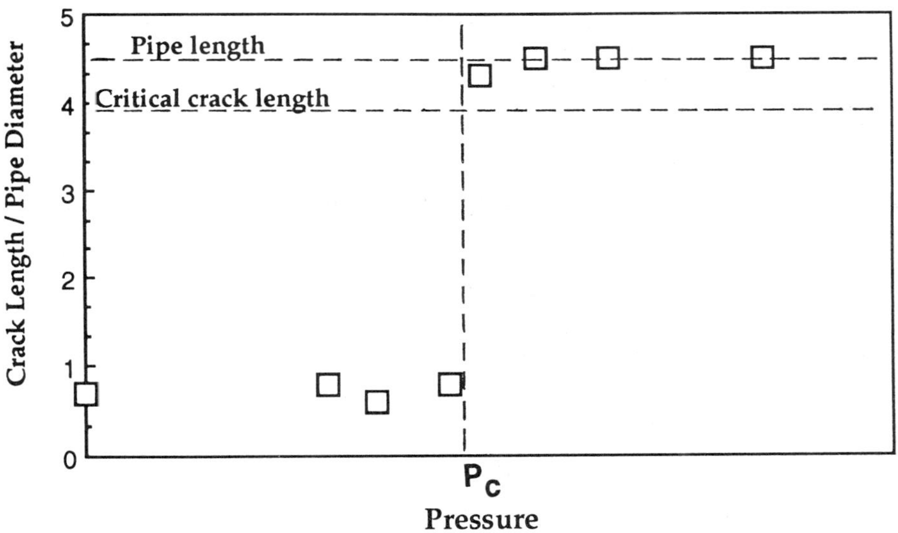

Fig. 1. Length of rapid fracture versus pressure in pipe. Typical data.

The elementary theory for the critical pressure, P_c, is based on the Irwin-Corten equation.

$$P_c = \frac{1}{SDR}\sqrt{\frac{8EG_F}{\pi D}} \tag{1}$$

where SDR is the ratio of pipe diameter over thickness; D is the pipe diameter and the material parameters are the modulus of elasticity, E, and the energy to produce the fractured surface, G_F. The Irwin-Corten theory indicates that the likelihood of RCP increases in accordance with the product P^2D where P is the pressure in the pipe and D is the diameter. There are two important experimental observations that are not taken into account by the Irwin-Corten theory; (1) as the pipe becomes thinner, there is a significant increase in P_c as shown by Leevers et. al. [4] which is not taken into account by equation (1); (2) the speed with which the crack is initiated affects the results. Consequently, the control of the impact speed of the striker has been mandated in the ISO specification for the S4 test.

The visual appearance of the fractured surface is also of interest. When RCP occurs almost all of the fractured surface is extremely smooth which is characteristic of very high speed fracture. However, the edges of the fractured surface are much rougher and indicate that the energy to produce fracture is greater at the edge of the surface than in the smooth interior surface. The reason for this difference between the interior and edges of the fractured surface is related to the fact that the interior surface was fractured under plane strain conditions and the edge under plane stress conditions. Although the percent of the fractured surface that contains the edge part is small (less than about 10%), the contribution of the edge to G_F in (equation 1) is dominant, because the specific fracture energy of the edge is much greater than the specific fracture energy of the smooth interior surface. For this reason, one can understand why P_c (equation 1) increases as the pipe thickness is decreased. The effect of specimen thickness on fracture toughness is a well known phenomenon as exhibited by the work of Mandell et. al. [5] on polyethylenes.

The importance of the edge of the surface in determining P_c was substantiated in an experiment by Leevers et. al. [4]. The high fracture energy of the inner edge was suppressed by making a 1 mm deep axial razor cut along the entire bore of the pipe. As a result, for a very tough pipe, P_c at -15°C dropped from 10 bar to 0.75 bar. According to the Irwin-Corten theory this decrease in P_c represents a decrease in the fracture energy, G_F, by a factor of 178.

In this paper the effect of loading rate on the initiation of RCP was investigated. It was found that there is a critical loading rate for initiating rapid fracture; however to produce long range rapid fracture as in RCP about a 50% higher loading rate was required. K_{IC} was nearly independent of loading rate. Increasing the temperature of the test makes it more difficult to produce RCP.

Experimental

The geometry of the test specimen is shown in Fig. 2. The specimen was taken from a 144 mm diameter medium density gas pipe. The notch was rough cut to a depth of 23 mm and then sharpened with a razor blade to a final depth of 25.4 ± 0.3 mm. The specimens were fractured under tensile loading with grips that conformed to the curvature of the specimen. The loading rates were varied from 22×10^4 to 93×10^4 N/s in a high speed servo hydraulic testing machine. There were two testing temperatures, 10°C and 28°C. In order to determine the loading rate, the load-time and deflection-time curves were recorded on an oscilloscope as shown in Figure 3.

The occurrence of RCP was readily observed by a visual inspection of the fractured surfaces as shown in Figure 4.

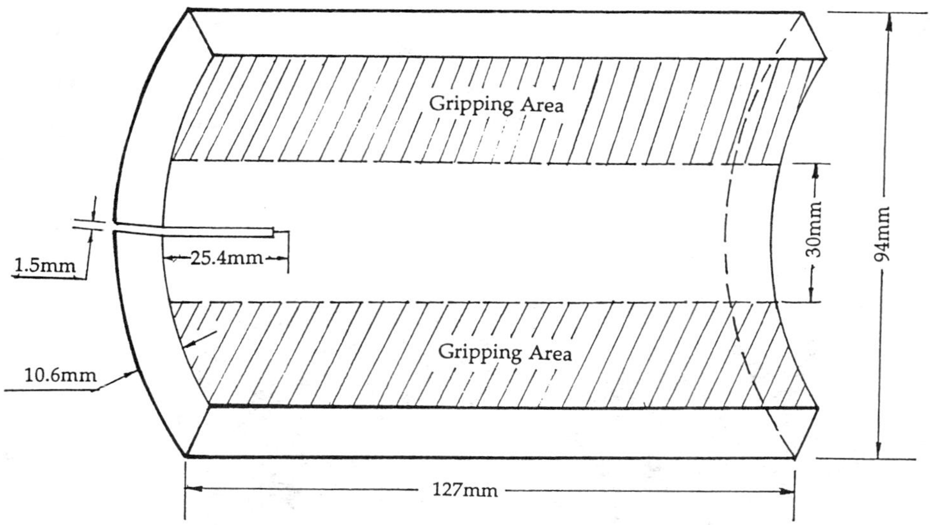

Fig. 2. Specimen geometry from 4" gas pipe.

Results

The results are shown in Table I for 10°C and 28°C tests. The loading rates were obtained from the initial slope of load-time curves as exhibited in Fig. 3.

The visual examination of the fractured surfaces as shown in Fig. 4 exhibited the following features. Immediately beyond the notch there is a white rough fractured surface which extends for a few mm or less. This fracture is a stable fracture of the original craze which formed during the loading of the specimen. It grows at a rate on the order of 0.5 m/s. When the crack growth becomes unstable, it reaches speeds of about 100 m/s and produces the very smooth fracture surface. If the unstable fracture reaches nearly to the end of the specimen, then RCP has occurred and would have continued over a greater distance if the specimen had been longer. The final rough fracture at the very end of the specimen is caused by the bending of the specimen in the remaining ligament beyond the region of RCP.

From the data in Table I a plot (Fig. 5) was made of the length of the rapidly fractured surface versus loading rate as 10°C and 28°C. At 10°C rapid fracture is initiated at a loading rate greater than 26×10^4 N/s. The length of the rapid fracture increases with loading rate and probably would increase indefinitely if the specimen was longer and if the loading rate was greater than about 40×10^4 N/s. The change in slope beyond 50 mm is associated with end effects as the specimen bends. When the length of rapid fracture is more than 50 mm which is 5 times the thickness, then it is expected RCP would proceed indefinitely as determined by the length of the specimen.

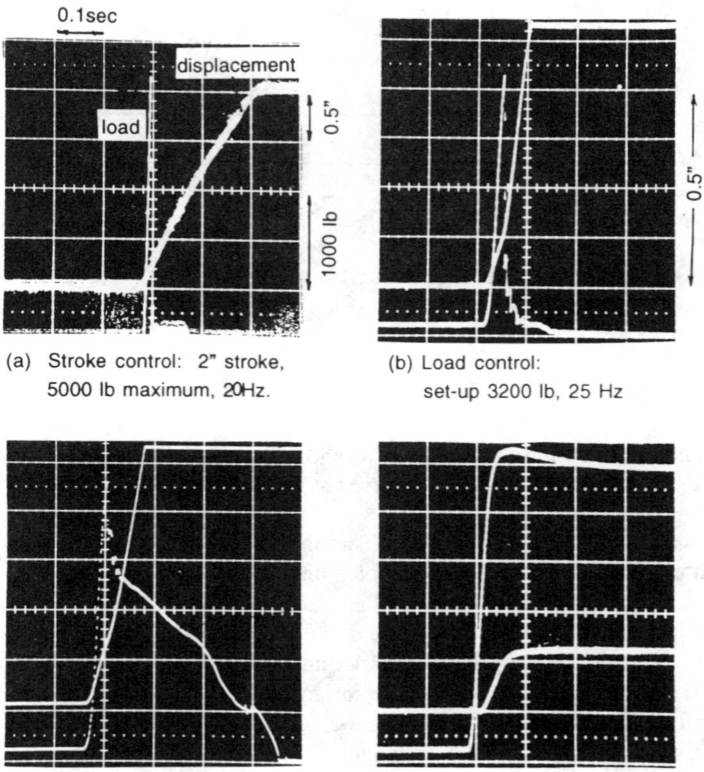

(a) Stroke control: 2" stroke, 5000 lb maximum, 20Hz.

(b) Load control: set-up 3200 lb, 25 Hz

(c) Load Control: set-up 3100 lb, 25 Hz

(d) Load control: set-up 3000 lb, 25 Hz

Fig. 3. Load-time curves and Displacement-time curves. (a) 84×10^4 N/s; (b) 42×10^4 N/s; (c) 35×10^4 N/s; (d) 24×10^4 N/s.

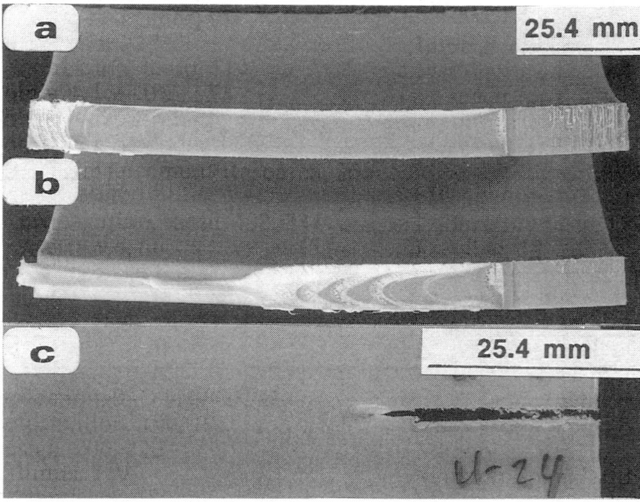

Fig. 4. Specimens fractured at 10°C at various loading rates (a) 93×10^4 N/s, long range rapid fracture. (b) 35×10^4 N/s, short range rapid fracture. (c) 23×10^4 N/s, no rapid fracture.

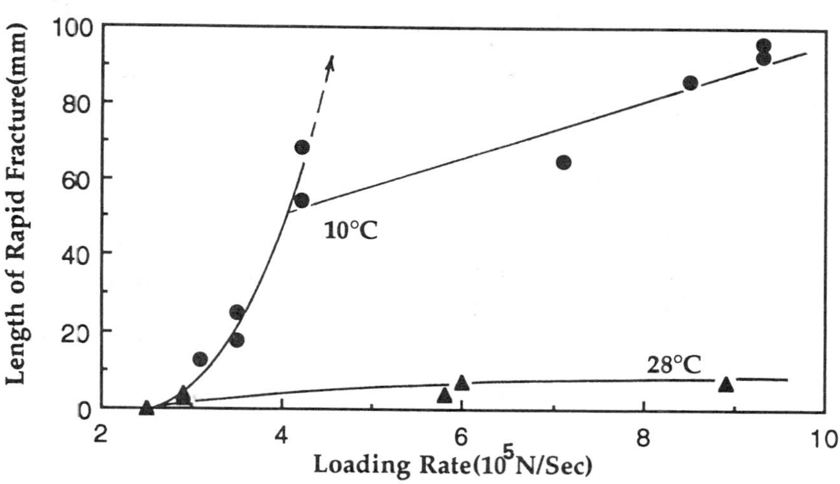

Fig. 5. Length of rapid fracture versus loading rate. Dotted line would correspond to an extremely long specimen.

Table I - Effect of Loading Rate on RCP and K_{IC} at 10°C and 28°C

10°C Tests

Specimen	Loading Rate (10^4 N/s)	Distance of Rapid Fracture (mm)	Maximum Load (10^3 N)	K_{IC} (MPa·m$^{1/2}$)
21	24	0	14.1	No Fracture
24	24	0	14.1	No Fracture
26	31	9	12.8	3.84
25	35	35	12.4	3.87
22	41	88	12.4	3.71
23	42	86	11.8	3.56
18	71	88	10.2	3.13
20	84	84	11.8	3.62
19	93	91	13.2	4.05
17	93	95	13.1	3.96
				av. 3.7±0.3

28°C Tests

Specimen	Loading Rate (10^4 N/s)	Distance of Rapid Fracture (mm)	Maximum Load (10^3 N)	K_{IC} (MPa·m$^{1/2}$)
8	22	0	10.9	No Fracture
11	29	4	10.9	3.29
9	29	3	10.9	3.29
10	58	4	11.6	3.63
12	60	6	10.9	3.29
27	88	7	10.2	3.08
				av. 3.3±0.2

The data at 28°C show that a small length of rapid fracture can be initiated at about a loading rate greater than 29×10^4 N/s, but even at a loading rate of 88×10^4 N/s, the rapid fracture could only travel about 5 mm which is only 1/2 times the pipe thickness. This result is expected because the energy for fracture greatly increases with temperature. Since 88×10^4 N/s was about the highest loading rate that can be achieved with our tensile machine, it is not known whether long rage brittle fracture can be produced in this specimen at 28°C.

Table I shows how the maximum load varies with loading rate. The maximum load coincides with the initiation of rapid fracture because beyond the maximum load the load very rapidly decreases. Therefore the maximum load can be converted to a K_{IC} which is defined as the stress intensity to initiate fracture where

$$K_{IC} = Y\sigma a^{1/2} \quad (2)$$

σ is the stress; a is the notch depth and Y is the geometric and loading factor. For our single edge notched specimen whose ratio of notch depth to specimen width is 0.2, Y = 2.43 as given by Williams [6].

For the loading rates from 31x10^4 to 93x10^4 N/s., K_{IC} was practically independent of loading rate with an average value of 3.7±0.3 $\text{MPa}\cdot\text{m}^{1/2}$ at 10°C and 3.3±0.2 $\text{MPa}\cdot\text{m}^{1/2}$ at 28°C. It is interesting to note that Barry and Delatycki [7] determined K_{IC} for a group of polyethylenes at 23°C under three point bending for strain rates from 2x10^{-7} s^{-1} to 0.1 s^{-1}. They found that over this range of strain rates K_{IC} increased by a factor of 3-4. In our experiments the strain rates were only varied by a factor of 4 so that our variation in strain rates was too small to produce a significant change in K_{IC}.

The input energy up to peak load was measured from the area under the curve of load versus time and from the displacement versus time curves. The input energy, when fracture was initiated, is given by

$$E = \frac{L_{max} D_{max}}{2} \quad (3)$$

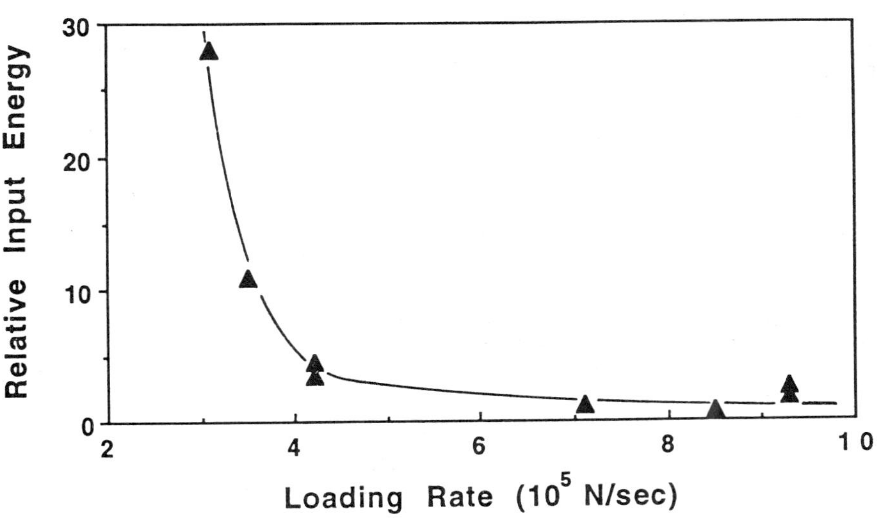

Fig. 6. Relative input energy at peak load versus laoding rate.

where L_{max} and D_{max} are the maximum load and the corresponding displacement. Fig. 6 shows the plot of E versus loading rate at 10°C. There is a rapid decrease in E when RCP occurs. Lee, et. al. [8] measured the fracture toughness and crack instability in tough polymers under plane strain conditions. They found that in general there was a decrease in the fracture energy when rapid fracture occurred. This decrease in energy was caused by a transition from multiple crazing to the single craze that is associated with rapid fracture.

Summary

The stress intensity, K_{IC}, to initiate fracture is independent of the loading rate. However the initiation of unstable rapid fracture occurs above a critical loading rate. To produce RCP, long range rapid fracture, requires a further increase of loading rate by about 50%. The input energy required to initiate fracture depends on loading rate where the energy to initiate rapid fracture is about ten times the energy required to produce long range rapid fracture. Thus a rapid low energy impact is more likely to produce RCP than a slow high energy impact. Whereas RCP can be produced at 0°C, it was not possible to produce RCP at 28°C with a high speed servo hydraulic machine for the particular gas pipe used in this investigation.

References

[1] Greig, J. M., "Rapid Crack Propagation in Hydrostatically Pressurized 250 mm Polyethylene Pipe", Paper 12, 7'th International Conference Plastic Pipes, Bath, England (19-22 September 1988).

[2] Yayla, P. and Leevers, P. S., "A Novel Technique for the Study of Rapid Crack Propagation In Plastic Pipes", 8'th European Conference on Fracture, Torino, Italy (1-5 October, 1990)

[3] Kanninen, M. F., Leung C.-P., Grigory S. C., Couque, H. R. and Popelar, C. F. "A Fracture Mechanics Methodology for Preventing Rapid Crack Propagation in PE Gas Distribution Pipes" Twelve Plastic Fuel Gas Pipe Symposium, Boston, Mass. (24-26 September 1991) p. 70.

[4] Leevers, P.S. and Yayla, P., "Thickness Effects on Rapid Crack Propagation in Polyethylene Pipe", ibid. p. 58.

[5] Mandell, J. F., Roberts S. R., McGarry, F. J., "Plane Strain Fracture Toughness of Polyethylene Pipe Materials" Polymer Engineering and Science, Vol. 23, (1983) 404.

[6] Williams, J. G., "Fracture Mechanics of Polymers", pub. John Wiley & Sons, New York (1984) p. 66.

[7] Barry, D. B. and Delatycki, O., "The Strain Rate Dependency of Fracture in Polyethylene: Fracture Initiation", J. of App. Poly. Sci., vol. 38, (1989) 339.

[8] Lee, L. H., Mandell, J. F., and McGarry, F. J., "Fracture Toughness and Crack Instability in Tough Polymers Under Plane Strain Conditions", Polymer Engineering and Science, 27 (1987) 128.

ACKNOWLEDGMENT

The research was sponsored by the Gas Research Institute. The central facilities as provided by the NSF-MRL grant DMR 91-20668 were most helpful. The assistance of Dr. Alex Radin was invaluable.

Author Index

Subject Index